College Arithmetic

3 1 2 4
6
9 5
7
8 0

W. I. LAYTON

Professor of Mathematics
and Head of Department
Stephen F. Austin State University
Nacogdoches, Texas

College

Arithmetic

Second Edition

JOHN WILEY & SONS, INC.
New York · London · Sydney · Toronto

Copyright © 1971, by John Wiley & Sons, Inc.

All rights reserved. Published simultaneously in Canada.

No part of this book may be reproduced by any means, nor transmitted, nor translated into a machine language without the written permission of the publisher.

Library of Congress Catalogue Card Number: 73-155121

ISBN 0-471-51976-6

Printed in the United States of America

10 9 8 7 6 5 4 3 2 1

To my wife, Eva Wade Layton

Preface

College Arithmetic, Second Edition, is intended primarily for students entering college without the mathematical understanding and skills necessary for adequate handling of the quantitative problems that arise in everyday affairs.

The text is largely a review of arithmetic, but topics from algebra, commercial arithmetic, and geometry are also included.

This new edition is an improvement over the original book in the following ways.

1. Many problems have been included that can be worked quickly.
2. All exercises are new.
3. The number of worked-out examples has been considerably increased.
4. The new edition contains more problems.
5. A new and more attractive format is used.
6. A significant number of simple and intermediate problems lead to the more involved problems.
7. Asterisks are used in this edition to indicate problems that are longer and more difficult and that should be assigned accordingly.

Word problems are mixed with drill. Many of the problems deal with personal budgeting, business, home management, hobbies, and leisure activities. Historical background is included for added interest and variety.

The greatest improvement over the first edition is in the exercise sets. In all topic areas there is a more than adequate selection of problems from which to choose. This new edition remains brief and yet covers a wide and quite comprehensive area.

Answers to the odd-numbered problems are included in the textbook. An instructor's manual is available containing answers to even-numbered problems. Sample tests with answers also appear in the manual.

Generally, teachers using this book probably will wish to consider the topics in the order in which they appear. Some teachers may prefer, however, to follow Chapter Four, "Decimals," with Chapter Seven, "Approximate Numbers."

In this second edition, I have presented elementary mathematics in such a way that, although little mathematical background is required of the student, he is aware of being treated as a mature individual. I sincerely hope that the student will become more enthusiastic about mathematics after having worked with this book.

This edition has the benefit of comments from both teachers and students who have used the first edition. I thank all who have contributed the many helpful suggestions.

I am deeply indebted to my wife for her patience, understanding, and valuable stenographic assistance in this project.

W. I. Layton

Contents

1 | Introduction 2

2 | Whole Numbers 6

3 | Fractions 32

4 | Decimals 64

5 | Weights and Measures 82

6 | Percentage 112

7 | Approximate Numbers 130

8 | Elementary Algebra 148

9 | Measurement of Plane and Solid Figures 178

Answers to Odd-Numbered Problems 213

Index 233

College Arithmetic

Chapter 1

Introduction

1 | Why study mathematics?

You will approach the study of arithmetic with mixed feelings. Very probably some of you have always liked mathematics but have been away from an active study of it for a number of years. Others of you may have felt that mathematics is simply "not for me." In either case, this textbook is intended to be a reorientation for you as far as the basic ideas of mathematics are concerned. It would be well to keep in mind that you should become more skilled in the handling of the elementary mathematics that is so essential in all vocations.

Let us consider briefly the impact of mathematics on past and present civilizations. Mathematics had its beginning in prehistoric times. Lancelot Hogben says, "The history of mathematics is the mirror of civilization."*

Many different countries have contributed to mathematics during its centuries of growth, and it has been a common heritage for much of mankind. Today, throughout the civilized world it is regarded as an absolute necessity for further progress.

Although much of mathematics is now highly refined, we are sure that its origin had to do with the everyday matters of food, clothing,

---

* Lancelot Hogben, *Mathematics for the Million*, W. W. Norton and Company, New York, p. 32, 1937.

and shelter. The basic questions, how much? how many? how long? could be answered only by counting and measuring. These processes under incessant pounding from hunger, cold, and desire gave the basic incentive for the creation of mathematics, and today, in spite of its tremendous growth, number and form remain among the fundamentals of the subject.

2 | How to study mathematics

Success in the study of mathematics, like success in the study of other subjects, comes as a result of enthusiastic application on the part of teacher and student. Keep an open-minded attitude and resolve to succeed in your study of mathematics. Go through each explanation in the textbook carefully with pencil and paper handy in order that you may delve into any steps you do not understand. Ask questions freely and remember that, although mathematics is not easy, it is not a subject that should be studied only by those who are the most highly gifted mentally. Intellectual curiosity and persistence are perhaps the main prerequisites for success. Make it a point not to leave a problem or an idea unless you really understand it.

3 | A glance at what lies ahead

In this book we shall cover the following topics. First, we shall review the fundamental operations of addition, subtraction, multiplication, and division with whole numbers. We shall apply these operations in solving problems. We shall introduce ways of checking our work in these operations, and shall also begin to look into the nature of our number system. Second, we shall take up fractions, and follow this topic with decimals and their applications in business problems and technical work. We shall turn to denominate numbers—the numbers we use in measuring quantities. We shall take up exponents and powers, which are very important in writing numbers as we use them in science and technology. Then we shall consider percentage and approximate numbers. After reviewing these fundamentals we shall do some work with elementary algebra, which is largely a generalization of principles in arithmetic, and we shall finish with some work in the geometry of measuring lengths, areas, and volumes.

Throughout the book we shall try to emphasize the review of elementary principles sufficiently to refresh your memory in the facts and ideas of arithmetic. We want to provide enough practice to enable you to master the ideas. Many of the problems to solve will be those met in everyday affairs in business, civil service work, consumer budgeting, and the basic phases of planning and management. Numerous problems of interest to technical and vocational students are also included. We hope to give you sufficient familiarity with the fundamentals of arithmetic to enable you to carry on successfully with further work in mathematics.

It is a long way from the arithmetic of the multiplication tables to the mathematics of modern technology, but the principles of the multiplication table still underlie a great deal of the work done in such tasks as computing satellite orbits. The first step in the direction of understanding and handling the mathematics of satellite orbits is a sound knowledge of elementary arithmetic. Beyond this step the training progresses through algebra, geometry, trigonometry, analytic geometry, calculus, differential equations, and other branches of mathematics. The experience common to those persons who have made the ascent is that the view from the upper levels is well worth the climb.

Those students who would like to study some of the ideas in higher mathematics and modern mathematics from an elementary point of view will find the following books interesting. Only a knowledge of arithmetic and an interest in numbers are prerequisite to the study of these books.

1. Bell, E. T., *Development of Mathematics*, New York, Simon and Schuster, Second Edition, 1945.
2. Bell, E. T., *Men of Mathematics*, New York, Simon and Schuster, 1937.
3. Dantzig, Tobias, *Number: The Language of Science*, New York, The Macmillan Company, Fourth Edition, 1954.
4. Kasner, E., and J. R. Newman, *Mathematics and the Imagination*, New York, Simon and Schuster, 1940.
5. Newman, James R., *The World of Mathematics*, Four Volumes, New York, Simon and Schuster, 1956.

Chapter 2

Whole Numbers

1 | Introduction

The numbers 1, 2, 3, 4, 5, 6, . . . are called the *natural numbers* because it is usually recognized that they have in some philosophical sense a natural existence independent of man. The most complicated number systems, on the other hand, are regarded as inventions of the mind of man. This infinite set of numbers has been represented in different ways in the course of history, another rather commonly known representation being the system of Roman numerals I, II, III, IV, V, VI, etc., which we shall discuss in this chapter. The most common representation, 1, 2, 3, . . . , is more often called the Hindu-Arabic or Indo-Arabic system because of the invention of this notation in India. Knowledge of this way of writing numbers was transmitted from India to the Western world by the Arabs. The natural numbers 1, 2, 3, 4, 5, 6, . . . are also known as whole numbers, or integers.

The notation for the natural numbers, 1, 2, 3, 4, 5, 6, 7, 8, 9, 10, 11, 12, 13, 14, . . . is called the *decimal system* (from the Latin *decem*, meaning ten) because ten symbols or digits are used to represent all numbers. This excellent system of notation is very compact since it is a place or positional system. Thus, in such a number as 632, the digit 6 represents 6 hundreds; the digit 3, 3 tens; and the digit 2, 2 units.

Referring to Figure 1, the digit *3* represents *three units*. A digit

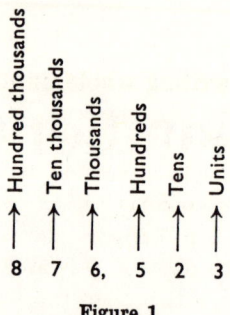

Figure 1

in this position is described as being in the *units place*. The digit 2 immediately to its left is in the *tens place* and represents 2 *tens*. Furthermore, the 5 is in the *hundreds place* and represents 5 *hundreds;* the 6 is in the *thousands place* and represents 6 *thousands;* the 7 is in the *ten thousands place* and represents 7 *ten thousands;* and the 8 is in the *hundred thousands place* and represents 8 *hundred thousands.*

Another interesting way to look at the number 876,523 is shown in Figure 2. By comparison, the Roman numeral XXX for 30 has three symbols X of equal value, 10. Let us notice that the Hindu-Arabic system can be used to represent numbers of any size with 10 symbols, whereas a non-positional notation such as the Romans used would require new symbols for reasonable brevity in the writing of much larger numbers.

The decimal system is said to have the *base ten*. This system probably originated from the fact that we have ten fingers for use in counting. Ten, however, is not the only possible base. We can use *any natural number greater than one*. In a later portion of this book we shall go into some discussion of the use of two as a base. We shall also elaborate on the meaning of ten as a base.

$$\begin{aligned}
&3 \text{ or } 3 \text{ times } 1 \text{ unit} \\
&20 \text{ or } 2 \text{ times } 10 \text{ units} \\
&500 \text{ or } 5 \text{ times } 100 \text{ units} \\
&6{,}000 \text{ or } 6 \text{ times } 1{,}000 \text{ units} \\
&70{,}000 \text{ or } 7 \text{ times } 10{,}000 \text{ units} \\
&800{,}000 \text{ or } 8 \text{ times } 100{,}000 \text{ units} \\
&\overline{} \\
&876{,}523
\end{aligned}$$

Figure 2

2 | Reading and writing whole numbers

It is quite important that we be able to read and write whole numbers. It is convenient to group the digits of a number and to separate these groups by commas. By digits we mean any of the integers 0, 1, 2, 3, 4, 5, 6, 7, 8, 9 appearing in a number. The number 362, for example, has the digits 3, 6, and 2. This number is read three hundred sixty-two, rather than three hundred and sixty-two as we perhaps might commonly read it. The use of *and* in three hundred and sixty-two is approved by the best authorities on English grammar, and its restriction, which we have just referred to, is purely a mathematical convention. In reading 5647, for instance, we would say five thousand six hundred forty-seven rather than five thousand, six hundred and forty-seven.

Another practice recommended in mathematics in reading and writing numbers is that only the compound words from twenty-one to ninety-nine are to be hyphenated. For example, 48 should be written forty-eight. The number 564,897,398,736,416 is described as

564	897	398	736	416
trillion	billion	million	thousand	units

This number is read five hundred sixty-four trillion, eight hundred ninety-seven billion, three hundred ninety-eight million, seven hundred thirty-six thousand, four hundred sixteen. It should also be noted that the words thousand, million, billion, etc., are written in the singular form. To read larger numbers than the one above requires a knowledge of words beyond trillion.

Example 1

Read the digit position that is underscored in 8728.
The 7 is in the hundreds position and represents 700.

Example 2

Read the digit positions that are underscored in 529,356.
The 9 is in the thousands position and represents 9000. The 2 is in the ten thousands position and represents 20,000. Thus the 29 in the illustration represents 29,000.

EXERCISE 1

Write the following numbers in words.

1. 38	2. 47
3. 84	4. 56
5. 68	6. 92
7. 35	8. 73
9. 927	10. 736
11. 347	12. 254
13. 590	14. 693
15. 806	16. 780
17. 8653	18. 4836
19. 7364	20. 4289
21. 3635	22. 5307
23. 8694	24. 6483
25. 2003	26. 5067
27. 8090	28. 6057
29. 5006	30. 8907
31. 34,617	32. 92,541
33. 19,657	34. 78,543
35. 29,760	36. 78,659
37. 87,423	38. 43,892
39. 536,853	40. 765,384
41. 3,589,462	42. 9,754,853
43. 45,392,762	44. 83,865,645
45. 745,632,429	46. 843,543,912
47. 68,435,629,036,259	48. 729,642,358,008,354
49. 437,812,947,638,217	50. 89,757,438,714,342

Read the digit positions that are underscored.

51. 7643	52. 8734
53. 6532	54. 3962
55. 43,653	56. 67,859

WHOLE NUMBERS ─────────────────────────────────── **11**

57. 83<u>9</u>,652   **58.** 4<u>28</u>,438
59. 6<u>54</u>,281   **60.** <u>837</u>,654

Write the following as numbers.

61. Six hundred forty-two

62. Nine hundred sixty-eight

63. Eight hundred thirty-five thousand, eight hundred seventy-two

64. Sixty-five million, eight hundred thirty-seven thousand, ninety-eight

65. Thirty-eight trillion, six hundred thirty-nine billion, eight million, sixty-five thousand, six hundred forty-two

66. Four hundred sixty-three billion, thirty-seven million, four hundred eighty-five

67. Nine hundred sixty-two billion, four hundred thirty-one million, eighty-seven thousand, nine

68. Eighty-three trillion, forty-seven billion, thirteen million, four thousand, five hundred thirty-seven

69. Four trillion, four billion, four million, four thousand, four hundred

70. Sixty-eight billion, forty-eight thousand, fifty-three

3 | Adding and subtracting whole numbers

In adding and subtracting whole numbers, as in the other phases of mathematics we shall come to, we need to concentrate in particular on accuracy and speed. Another point to bear in mind is that we can add and subtract only numbers that are expressed in the same kind of unit. Thus we may add 16 tons and 23 tons, but we cannot add 25 tons and 18 quarts.

Addition is the process of finding the total number of items of a given kind in several sets of items of the same kind. In other words, 5 nickels and 2 quarters are equivalent to 1 quarter and 2 quarters, and their sum is 3 quarters. Also, if we are considering dimes and half dollars, 1 half dollar is equivalent to 50 cents and 1 dime is equal to 10 cents. Then upon adding 8 dimes and 3 half dollars, we obtain 80 cents plus 150 cents = 230 cents.

We use place value in our number system. This means that *10 units* are equal to *1 ten*, *10 tens* are equal to *1 hundred*, *10 hundreds* are equal to *1 thousand*, and so on. When we have totaled 10 in adding the items in one column, this is equivalent to 1 in the column to its left. We speak of *carrying 10* from one column to obtain *1* in the next column to the left.

For example, add:

$$\left.\begin{array}{r}36\\5679\\247\end{array}\right\} \text{ addends}$$

$$\overline{5962} \quad \text{sum}$$

The numbers to be added, 36, 5679, and 247, are called *addends*. The result of the addition, 5962, in the foregoing example, is called the *sum*.

In this example we notice that the numbers 6, 9, and 7, which are in the units place, are in one column. The numbers 3, 7, and 4, which are in the tens place, are in one column. The numbers 6 and 2, which are in the hundreds place, are in one column, etc. It is wise to keep these columns in clean-cut straight lines, as this organization helps to encourage accuracy.

We have pointed out in a recent paragraph that we can add and subtract only numbers that are expressed in the same kind of unit. It does not seem unreasonable therefore that units must be added to units, tens to tens, hundreds to hundreds, etc.

Subtraction is the process of finding the difference between the number of items of a given kind contained in one set, and the number of items of the same kind in a second set. The *difference* or *answer* in subtraction may also be considered as the number of items that *remain* in one set after a number of these items are removed from it.

Subtraction is the *inverse* of *addition*. By this statement we mean that if we add a number and then subtract the same number, the result is the original number. Thus:

$$18 + 5 = 23 \quad \text{and} \quad 23 - 5 = 18$$

Place value is also important in subtraction. We refer to *borrowing* one item from one column and converting it to ten items in the column to its right. In subtracting, of course, units are subtracted from units, tens from tens, hundreds from hundreds, etc.

In subtracting we also observe a careful lining up of the columns.

WHOLE NUMBERS

Thus in subtracting 25,642 from 95,429 we have

$$\begin{array}{ll} 95{,}429 & \text{minuend} \\ 25{,}642 & \text{subtrahend} \\ \hline 69{,}787 & \text{difference or remainder} \end{array}$$

In this example, 95,429 is called the *minuend;* 25,642 the *subtrahend;* and 69,787 the *difference* or *remainder.* Thus the number we are subtracting from is called the minuend, the number we are subtracting is called the subtrahend, and the result of the subtraction is known as the difference or remainder. Another way to look at the example is to say that, when we subtract 25,642 from 95,429, we have 25,642 items in one set to be subtracted from 95,429 items of the same kind in another set. The *difference* or *answer*, 69,787, is the number of items that remain in the set that contained 95,429 items, after a set that contains 25,642 items is removed or subtracted. The process of subtraction may be explained by the *decomposition* method or by the *equal additions* method.

We illustrate the *decomposition* method as follows.

Subtract 674 from 935.

$$\begin{array}{ccc} \text{hundreds} & \text{tens} & \text{units} \\ 8 & 1 & \\ \cancel{9} & 3 & 5 \\ 6 & 7 & 4 \\ \hline 2 & 6 & 1 \end{array}$$

In the units column, 4 from 5 = 1. Write 1 in the units place in the remainder.

In the tens column we cannot take 7 from 3, so we change, or decompose, the 9 hundreds in the minuend into 8 hundreds plus 1 hundred. This 1 hundred is now changed to 10 tens (sometimes this is referred to as "borrowing") and added to the already present 3 tens, making 13 tens in all. It is now possible to subtract the 7 tens of the subtrahend from the 13 tens of the minuend. This leaves 6 tens. Write 6 in the tens place in the remainder.

In the hundreds column, 6 from 8 = 2. We write 2 in the hundreds place in the remainder.

Let us now examine the *equal additions method*.

```
         hundreds  tens  units
                    1
            9       3      5
            7
            6̸      7      4
           ─────────────────
            2       6      1
```

In the units column, 4 from 5 = 1. Write 1 in the units place in the remainder.

In the tens column we cannot take 7 from 3. We now add 1 hundred to both the minuend and the subtrahend. We see that when we add the same quantity to both minuend and subtrahend, we do not change the remainder. The 1 hundred is added to the minuend as 10 tens, giving 13 tens in all, and added to the subtrahend as 1 hundred, giving 7 hundreds in all. Now we can subtract the 7 tens in the subtrahend from the 13 tens in the minuend. This gives 6 tens. Write 6 in the tens place in the remainder.

In the hundreds column, 7 from 9 = 2. Write 2 in the hundreds place in the remainder.

The *addition check* is the basis for the most frequent check for subtraction. This check is based on the fact that the *remainder* of a subtraction can be thought of as a missing *addend*.

In checking subtraction by addition, add the remainder to the subtrahend. Now see if this sum equals the minuend of the problem.

Check the subtraction 984 − 438 = 546 by addition.

Adding the remainder to the subtrahend, we have

$$\begin{array}{r} 546 \\ 438 \\ \hline 984 \end{array}$$

Compare the sum 984 just obtained with the minuend 984 of the original problem.

WHOLE NUMBERS ─────────────────────────── 15

EXERCISE 2

Add in each of the following.

1. 3 2 8 1 0 7 6 8 9 8
    1 5 7 9 4 3 6 9 0 7
    ─ ─ ─ ─ ─ ─ ─ ─ ─ ─

2. 3 9 3 6 4 8 5 0 5 7
    6 5 0 8 4 7 4 8 2 3
    ─ ─ ─ ─ ─ ─ ─ ─ ─ ─

3. Add each column, first adding downward, second adding upward to check.

46	14	95	57	90	93
29	93	48	36	17	46
15	47	33	62	49	84
63	53	29	70	88	36
37	88	47	33	96	77
66	30	32	66	45	60
91	14	68	71	79	88
40	77	49	98	28	42
──	──	──	──	──	──

4. Add each column downward, then add upward to check.

473	957	196	807
608	83	394	555
999	506	683	712
472	717	207	976
396	84	806	388
884	668	978	403
───	───	───	───

Add in each of the following.

5. | 7384 | 4694 | 7964 | 7964 | 8659 |
    | 2657 | 5268 | 9638 | 853 | 4364 |
    | ──── | ──── | ──── | ──── | ──── |

6.

7	36	629	7638	55,496
4	59	37	4317	334
9	42	5	9257	4,965
6	38	684	2318	39,394
3	65	79	4563	87
2	97	357	7854	5,639
8	48	49	7964	85,784

7.

7	17	594	836	7846
8	6	486	78	4654
2	39	358	594	3476
9	8	593	6	5973
5	54	685	37	7864
4	33	207	794	3759

8.
6753
8469
5957
4836
2743

9.
38,579
5,436
69,758
83,623
68,416

10. Add downward; check by adding upward.

$392.83	$166.35
46.08	471.89
777.20	78.47
600.90	618.00
752.35	83.25
95.40	970.40

11. Add downward; check by adding upward.

$468.73	$6284.70
532.07	3207.50
27.86	2236.60
536.40	788.20
875.70	9650.00
326.28	7000.40

WHOLE NUMBERS

12. Add each set of numbers horizontally, without rewriting, and reading from left to right.

$34 + 65 + 78 + 19 + 52 + 66 + 47 + 69 + 55 =$
$58 + 33 + 17 + 82 + 44 + 63 + 79 + 54 + 31 =$
$12 + 63 + 47 + 80 + 93 + 57 + 49 + 68 + 42 =$
$47 + 60 + 74 + 16 + 37 + 45 + 68 + 94 + 15 =$
$\$46.50 + \$17.35 + \$18.40 + \$67.26 + \$94.83 = \$$
$\$15.62 + \$47.91 + \$35.62 + \$84.51 + \$46.14 = \$$

Check each problem by adding horizontally from right to left.

Add the following:

13.
347
97,623
44
609,395
549
76,854
793,568
9,483,672
34,869,547
40,807

14.
5,368
764,782
43,658
8,888,888
35,692
384,473
84,697
49,457,788
42
348,694

15.
45,964
835,627
36,755
949,652
97,643,846
2,584,892,654
38,553
684,357,765
638,446
5,257,694

Subtract:

16. 38   46   24   56   68
     5   8   9   7   6

17. 25   34   87   92   87
     4   7   9   5   8

18. 47   63   84   93   75
     28   35   76   84   69
     ——   ——   ——   ——   ——

19. 576   856   743   329   802
     324   249   568   108   478
     ——   ——   ——   ——   ——

20. 856   763   976   473   680
     437   588   465   368   476
     ——   ——   ——   ——   ——

21. 9837   7329   4803   6483   8459
     4786   5693   2961   3296   5678
     ——   ——   ——   ——   ——

22. 7842   6788   8658   5080   7496
     5937   3986   4793   3567   4879
     ——   ——   ——   ——   ——

Subtract in each of the following, and check by means of the *addition check*.

23. 7946
     399
     ——

24. 49,484
     25,377
     ——

25. 785,637
     543,668
     ———

26. 8,629,427
     866,999
     ———

27. 7,643,224
     3,527,555
     ———

28. 7004
     5038
     ——

29. 89,004
     7,806
     ——

30. 7,692,334
     4,676,555
     ———

31. 347,326,400,007
     81,017,309,888
     ————

32. 85,796,300,007
     8,085,897,693
     ————

33. How many pennies would there be in a collection equivalent to the following? 5 quarters, 8 dimes, 3 nickels, and 20 pennies. How many nickels would there be in a collection having the same value?

WHOLE NUMBERS   19

34. Tom has 3 quarters, 5 nickels, and 8 pennies. John has 5 quarters and 2 nickels. Which one has more money, and how much more does he have?

35. Mayor Johnson received 8126 votes, and Mr. James received 6439 votes. How many more votes did Mayor Johnson receive?

36. There are 142 boys and 158 girls in the freshman class at the Wells High School, 124 boys and 136 girls in the sophomore class, 175 boys and 135 girls in the junior class, and 112 boys and 164 girls in the senior class. How many boys are enrolled at the school? How many girls? What is the total enrollment?

37. How many hours will a program last if there are portions taking three-quarters of an hour, 6 minutes, 12 minutes, 9 minutes, 18 minutes, and one-half hour?

38. John has saved $49.67. He wants to buy a suit selling for $67.50. How much more does he need?

39. Mr Smith deposited the following amounts in his bank account during a period of one month: $108.53, $62.79, and $29.46. He wrote checks for the following amounts: $9.84, $4.236, $7.24, $19.39, $24.46, $15.50, $21.23, $69.75, and $121.69. If his original balance was $287.64, what was his balance at the end of the month?

40. The goal in a fund-raising drive is $48,360. If funds in amount of $32,845.63 have been raised, how much more is needed to reach the goal?

41. Pieces of baggage on a plane weighed 28 pounds, 32 pounds, 47 pounds, 51 pounds, 23 pounds, 42 pounds, and 29 pounds. What was the total weight of baggage?

42. On a trip a family drove distances of 368 miles, 470 miles, 272 miles, 596 miles, and 438 miles. They spent $29.68, $32.46, $39.60, $27.60, and $45.75. How far did they go, and how much did they spend?

43. On a certain construction project, 70 tons of concrete are needed. The first day 6 truckloads were used weighing respectively 2689, 3575, 13,689, 25,867, 4993, and 56,896 pounds. How many more pounds are needed to complete the project? (One ton = 2000 pounds.)

44. A television technician, Jake Smith, finds that his jobs and

sales for one day are as follows:
  MR. ATWOOD: Labor $4.50, parts $9.20.
  MR. ROSE: Labor $6.30.
  MR. SELF: Parts $14.68.
  MR. TUCKER: Labor $8.75, parts $34.42.
  MR. AMES: Labor $19.55, parts $24.19.
  (a) What were his receipts for the day?
  (b) How much of the receipts was for labor?

4 | Multiplying whole numbers

Multiplication is a process of finding the total of a number of equal sets. Thus the cost of 8 three-cent stamps is equivalent to 8×3 cents or 24 cents. The amount of pay received for 40 hours of work at $4.00 per hour is $40 \times (\$4.00)$, or $160.00. Multiplication is a short way of working out an addition problem in which a number of sets of items are added, each set having the same number of items. For example:

$$5 \times (14¢) = 14¢ + 14¢ + 14¢ + 14¢ + 14¢ = 70¢$$

Place value should be understood in multiplication. Thus:

$$\begin{array}{c} 342 \\ \times 63 \end{array} \quad \text{is equal to} \quad \begin{array}{c} 342 \\ \times (60 + 3) \end{array}$$

or

$$\begin{array}{cc} 342 & 342 \\ + & \\ \times 60 & \times 3 \end{array} \quad \text{is equal to} \quad \begin{array}{cc} 342 & 342 \\ + & \\ \times 60 & \times 3 \\ \hline 20{,}520 & 1026 \end{array}$$

equals

$$\begin{array}{r} 342 \\ \times 63 \\ \hline 1026 \\ 2052 \\ \hline 21{,}546 \end{array} \quad \begin{array}{l} \text{multiplicand} \\ \text{multiplier} \end{array}$$

In multiplying whole numbers, special attention should be given to accuracy and speed. In setting up numbers for multiplication we

WHOLE NUMBERS _____21

place units under units, tens under tens, hundreds under hundreds, etc., just as we did in addition and subtraction. For example:

$$\begin{array}{rl} 2643 & \text{multiplicand} \\ 679 & \text{multiplier} \\ \hline 23787 & \\ 18501 & \\ 15858 & \\ \hline 1{,}794{,}597 & \text{product} \end{array}$$

In this example, the 679 is called the *multiplier*, the 2643 is known as the *multiplicand*, and the 1,794,597 is referred to as the *product*. Thus the number that is to be multiplied is called the multiplicand. The number that shows how many times the multiplicand is to be taken is called the multiplier. The result obtained by multiplying two numbers is called the product.

EXERCISE 3

Multiply:

1. 7   9   4   8   9   2   7   4   5   0
    3   1   6   5   6   5   8   0   5   7
    —   —   —   —   —   —   —   —   —   —

2. 1   8   6   7   9   7   0   8   3   4
    8   5   3   8   3   5   4   8   0   1
    —   —   —   —   —   —   —   —   —   —

3. 46   18   23   44   59
     2   7   2   3   4
    —   —   —   —   —

4. 97   84   65   84   76
     3   2   3   4   6
    —   —   —   —   —

5. 24   37   57   86   75
    15   26   33   42   39
    —   —   —   —   —

6. 75   49   57   89   68
     23   38   63   47   87
     ──   ──   ──   ──   ──

7. 247   **8.** 796   **9.** 948
       83   73   67
      ───   ───   ───

10. 653   **11.** 763   **12.** 983
        842   897   729
       ───   ───   ───

13. 9846   **14.** 7847   **15.** 547
         397   685   8632
       ────   ────   ────

16. 658   **17.** 7849   **18.** 5437
        8793   8996   6892
       ────   ────   ────

19. 59,724   **20.** 87,629   **21.** 92,607
           603   593   7,598
       ──────   ──────   ──────

22. 347,896
            785
       ───────

23. Write in words the numbers to be multiplied together in Problem 21.

24. Write in words the numbers to be multiplied together in Problem 22.

Multiply:

*25. 865,423,659   ***26.** 965,387,967
              68,597   763,572
         ──────────   ──────────

27. Find the cost of each group of items:
    (a) Eighteen dish towels at 36 cents each.
    (b) Five dozen dishcloths at two for 25 cents.
    (c) Forty-six feet of chain at 34 cents per foot.

28. What are the gross earnings of a clerk who works 40 hours at $2.12 per hour, 8 hours at $3.18 per hour, and 9 hours at $4.24 per hour?

WHOLE NUMBERS 23

29. Find the cost of 85 four-cent stamps, 26 five-cent stamps, and 48 six-cent stamps.

30. How many square feet of window area are there in a wall that has 7 windows each measuring 3 feet by 7 feet?

31. How many cubic feet of air does a room 14 feet by 21 feet by 9 feet contain?

32. What is the cost of equipping a corps of 80 men with uniforms at $102.30 each?

33. How many pounds of food are needed to feed 540 men for 18 days, if each man eats 8 pounds of food per day?

34. What is the cost of 12 jet planes if each costs $24,869,600?

35. How far does a car travel in 8 minutes if it goes at a rate of 88 feet per second? (Hint: Distance = rate × time.)

36. What is the cost of 12 three-cent stamps, 45 five-cent stamps, and 30 six-cent stamps?

37. One roofer can put on 42 sq ft of cedar shingles per hour. How many square feet of roof can be shingled by 5 men in 8 hours if each man works at the stated rate?

38. A student deposits $84 in the bank each week of his summer vacation. If his summer vacation is 14 weeks long, how much has he deposited in the bank at the end of the summer?

39. A manufacturer ships 8 cases of buttons. Each case contains 1 gross (144) of cards of buttons. In 5 of the cases, the cards have 6 buttons on each card; in the other 3 cases, the cards have 4 buttons on each card. How many buttons were shipped?

5 | Dividing whole numbers

In dividing whole numbers, the number that we are dividing, called the *dividend*, and the number that we are dividing by, called the *divisor*, are placed in a special form. The answer is called the *quotient*, and the remainder, if there is one other than zero, is written over the divisor beside the quotient. *Division* is the inverse of multiplication. By this statement we mean that if we multiply by a number and then divide the result by the same number, the final result is the original number. Thus:

$$18 \times 5 = 90 \quad \text{and} \quad 90 \div 5 = 18$$

Consider now the division of 6784 by 97.

$$\begin{array}{r} 69 \\ \text{divisor } 97\overline{)6784} \\ 582 \\ \hline 964 \\ 873 \\ \hline 91 \end{array}$$

quotient
dividend

remainder

The complete result of the division is $69\frac{91}{97}$. To check this result multiply the quotient by the divisor and add the remainder. Thus:

$$\begin{array}{r} 69 \\ \times 97 \\ \hline 483 \\ 621 \\ \hline 6693 \end{array} \qquad \begin{array}{r} 6693 \\ +91 \\ \hline 6784 \end{array}$$

Division may be considered in at least two different ways.

First: Division is the process of measuring to find how many times one number is contained in another. The procedure may be carried out by repeated subtraction or by actual measurement. Thus, 8 inches would be contained in 53 inches 6 times, with a remainder of 5 inches. We find therefore that $8\overline{)53} = 6\frac{5}{8}$.

Second: Division is a process of partitioning or separating a quantity into a number of equal parts. Thus, to divide 5 into 35 is the same as to find one-fifth of 35. To divide a quantity by 5 is to find one-fifth of it. If 5 boys share 35 cents, each one receives one-fifth of 35 cents, or 7 cents.

EXERCISE 4

Divide and check.

1. $15\overline{)75}$ $24\overline{)96}$ $29\overline{)174}$ $35\overline{)245}$ $53\overline{)424}$
2. $26\overline{)624}$ $17\overline{)646}$ $19\overline{)342}$ $46\overline{)966}$ $33\overline{)462}$
3. $18\overline{)882}$ $23\overline{)851}$ $45\overline{)1755}$ $27\overline{)1269}$ $39\overline{)1014}$

WHOLE NUMBERS _____ 25

4. $38\overline{)1026}$ $23\overline{)966}$ $85\overline{)1360}$ $37\overline{)1554}$ $36\overline{)1044}$ $43\overline{)1591}$
5. $85\overline{)596}$ 6. $63\overline{)842}$
7. $43\overline{)927}$ 8. $58\overline{)725}$
9. $96\overline{)1543}$ 10. $24\overline{)3546}$
11. $123\overline{)7503}$ 12. $187\overline{)4301}$
13. $215\overline{)7955}$ 14. $364\overline{)16,744}$
15. $473\overline{)26,015}$ 16. $168\overline{)7056}$
17. $796\overline{)19,104}$ 18. $492\overline{)42,804}$
19. $268\overline{)15,812}$ 20. $684\overline{)31,464}$
21. $104\overline{)10,007}$ 22. $238\overline{)74,932}$
23. $756\overline{)35,472}$ 24. $837\overline{)67,543}$
25. $439\overline{)112,563}$ 26. $693\overline{)479,604}$
27. $9075\overline{)673,212}$ 28. $8437\overline{)747,819}$
29. $4216\overline{)936,714}$ 30. $5748\overline{)643,814}$
31. $38,642\overline{)745,936}$ 32. $87,948\overline{)957,842}$
33. $23,473,692\overline{)86,499,372}$ 34. $49,647,388\overline{)93,477,864}$

35. How many strips of wood 16 inches long can be cut from a board 224 inches long?

36. How many inches long would each of 8 equal parts of a board 224 inches long have to be?

37. How many miles per gallon does a car travel if it travels 798 miles on 42 gallons of gasoline?

38. How many cars 16 feet long and placed end to end are in a traffic jam 5 miles long? (5280 feet = 1 mile.)

39. How many dollars does each of 324 workers receive as a bonus, if each receives an equal share of $138,024?

40. A corporation declares an extra dividend of $842,563.46 on its 536,392 shares of stock. What is the extra dividend per share?

41. Two cities 432 miles apart are to be plotted on a poster with a scale of 1 inch = 18 miles. How many inches apart should they be plotted?

42. Mr. Jackson earns an annual salary of $9120. What is his monthly salary?

43. How many cars 16 feet long can be placed end to end along 3 miles of pavement?

44. A ship traveled a distance of 1216 nautical miles at a speed of 19 knots (1 knot = 1 nautical mile per hour). How many hours did the voyage take?

6 | The Roman notation for numbers

There was a time in Europe when the Roman system of notation was the most popular method of writing numbers. The system is still used rather frequently today for recording dates, numbering chapters in books, and for artistic purposes. The symbols we use today for writing numbers in the Roman notation are given in Table 1.

TABLE 1

Our Number	Roman Symbol
1	I
5	V
10	X
50	L
100	C
500	D
1000	M

Three principles are used to form numbers with Roman numerals: (1) addition, (2) multiplication, and (3) subtraction.

The principle of addition is illustrated as follows:

$$XXVIII = X + X + V + I + I + I = 28$$

The principle of multiplication means that drawing a bar over a number multiples that number by 1000.

For example:

$$\overline{XXVIII} = 28{,}000; \quad \overline{CCC} = 300{,}000$$

The principle of subtraction is used in writing all fours and nines. It is not used in any other numbers.

WHOLE NUMBERS

For instance:

$$4 = IV, \quad 9 = IX, \quad 40 = XL, \quad 90 = XC,$$
$$400 = CD, \quad 900 = CM$$

In writing numbers with Roman numerals, it is recommended that the following rule be followed: First write the thousands, then the hundreds, then the tens, and last of all, the units.

Following this rule we write 87 as LXXXVII and not as XIIIC. Also 632 would be written as DCXXXII and not as CCCCMXXXII.

Other illustrations of numbers correctly written by the Roman system are:

$$3264 = MMMCCLXIV \quad \text{or} \quad \overline{III}CCLXIV$$
$$29{,}897 = \overline{XXIX}DCCCXCVII$$
$$648{,}753 = \overline{DCXLVIII}DCCLIII$$

EXERCISE 5

Write the following in the Roman notation.

1. 8
2. 16
3. 25
4. 39
5. 42
6. 69
7. 94
8. 58
9. 87
10. 73
11. 126
12. 354
13. 859
14. 3647
15. 9768
16. 32,684
17. 386,842
18. 743,849

Write the following according to our method of notation.

19. XXV
20. XIV
21. XXXII
22. LXIII
23. XLVII
24. LXXXII
25. XLIV
26. LXXVII
27. XCIX
28. CXXV
29. CCCXXVII
30. DCCXXII
31. CDLVIII
32. MCCCXLVII
33. $\overline{VII}$CDLVIII
34. $\overline{XXIII}$CCCXXII
35. $\overline{CDLVI}$DCCXLIV
36. $\overline{LXII}$

EXERCISE 6

Review

Write the following numbers in words.
1. 7649
2. 14,386
3. 65,392,467
4. 832,658,529
5. 47,629,594,826

6. A state collects $849,625,927 in taxes in a certain year. Write the amount of taxes in words.

7. A company has a gross income of $736,543,628 in a certain year. Write this income in words.

8. A fraternity has 25 gallons of wine in a barrel. If 3 gallons are taken out each day and 2 gallons put in at night, in how many days will the barrel be empty?

9. A farmer counted his eggs by twos, threes, fours, fives, and sixes and had one left each time. He then counted them by sevens and none was left. What is the least number he could have had?

10. Write in numbers: four billion, seven hundred forty-three million, five hundred sixteen thousand, two hundred thirty-eight.

11. Write in numbers: eighteen trillion, nine hundred sixty-eight billion, four hundred thirteen million, three hundred twenty-seven thousand, seven hundred seven.

12. Add:
6539
704
4276
30
827

13. Add:
8647
139
8007
6932
704

14. Multiply:
73,615
9,049

15. Multiply:
629,814
6,079

WHOLE NUMBERS

16. Table 2 is the monthly expense report for the sales department of the Soft Leather Shoe Manufacturing Company for the month of August 1968. Complete the report by finding the totals as indicated.

TABLE 2

Salesman	Automobile Gas and Oil	Other	Hotel	Meals	Entertainment of Customers	Totals
Asper	$93	$34	$104	$174	$86	
Bond	98	39	112	185	95	
Colter	107	45	124	192	103	
Doo	102	43	108	169	98	
Evans	130	61	136	198	94	
Moore	114	42	115	147	89	
Totals						

17. The sales made by the salesman for the Eat-Mor Ice Cream Company for the week of June 12 through 17 are given in Table 3. Complete the report.

18. A flour mill averages 4692 barrels a day. (One barrel = 196 pounds of flour.) On the average, how many 25-pound sacks can be filled from one day's output?

TABLE 3

Salesman	Monday	Tuesday	Wednesday	Thursday	Friday	Saturday	Totals
Evers	$187	$216	$342	$398	$422	$486	
Freeman	246	108	487	346	129	159	
Howell	369	221	292	487	324	284	
Jarman	432	454	486	495	512	216	
Keown	218	146	361	272	319	283	
Totals							

19. Campbell works in a machine shop and has made the following castings: 420 pounds, 280 pounds, 360 pounds, 49 pounds, and 134 pounds. What will the castings be worth at 17 cents per pound?

Write the following numbers in words.

20. 72,628,542,069   **21.** 865,492,539,927

22. Marty has saved $53.67 and wants to buy a coat marked down to $64.99. How much more does she need?

23. The goal in a fund raising compaign is $104,620. If $96,709.38 has been raised, how much more is needed to reach the goal?

24. Find the gross earnings of a man who works 40 hours at $2.30 pet hour, 7 hours at $3.50 per hour, and 5 hours at $4.60 per hour.

25. How many inches long would each of 16 equal parts of a board 240 inches long have to be?

26. How many dollars does each of 849 workers receive as a bonus, if each receives an equal share of $353,184?

Chapter 3

Fractions

1 | Adding and subtracting proper fractions

A fraction represents one or more of the equal parts of a unit. Thus, $\frac{5}{8}$ inch represents the length obtained if 1 inch is divided into 8 equal parts and 5 of those parts are taken (Figure 3).

In the number $\frac{5}{8}$, the 8 gives the fraction the name eighths, and is called the *denominator;* the 5 states the number of equal parts that are taken, and is called the *numerator*. The numerator and denominator referred to together are called the *terms* of the fraction. Fractions that have no restrictions on the size of the terms are called common fractions. A simple fraction has integral terms, that is, terms that are whole numbers. A simple fraction is called proper if the numerator is less than the denominator; otherwise it is called improper. Thus $\frac{3}{8}$ is a proper fraction, whereas $\frac{5}{4}$ is an improper fraction. Similarly, $\frac{6}{6}$ is an improper fraction, while $\frac{5}{6}$ is a proper fraction.

Equivalent fractions are fractions that *represent the same portion or part of a whole quantity*. Thus, $\frac{5}{8}$ of 1 inch is equivalent to $\frac{10}{16}$ of 1 inch. Five-eighths of 1 inch is equivalent to $\frac{20}{32}$ of 1 inch. Five-eighths of 1 inch is equivalent to $\frac{25}{40}$ of 1 inch. Also, $\frac{200}{400}$ of 1 dollar is equivalent to $\frac{1}{2}$ of 1 dollar, and $\frac{50}{100}$ of 1 dollar is equivalent to $\frac{1}{2}$ of 1 dollar.

Equivalent fractions can always be changed to identical fractions either by multiplying both terms of one fraction by the same number or by dividing both terms of one fraction by a suitable number. For example:

$$\frac{18}{27} = \frac{18 \div 9}{27 \div 9} = \frac{2}{3} \text{ ; and } \frac{35}{63} = \frac{35 \div 7}{63 \div 7} = \frac{5}{9}$$

Here we have reduced the fractions $\frac{18}{27}$ and $\frac{35}{63}$. Also consider:

$$\frac{2}{5} = \frac{2 \times 20}{5 \times 20} = \frac{40}{100} \text{ and } \frac{3}{8} = \frac{3 \times 45}{8 \times 45} = \frac{135}{360}$$

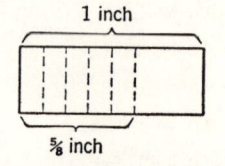

Figure 3

Here we have changed the fractions $\frac{2}{5}$ and $\frac{3}{8}$ into equivalent fractions containing larger numbers. In this paragraph we have seen in operation the *Fundamental Principle of Fractions*, which states that multiplying or dividing both terms of a fraction by the same number does not change its value.

We say that $\frac{18}{27}$ has been *reduced to lower terms* and that $\frac{2}{5}$ has been *changed to higher terms*. A simple fraction may be changed to higher terms by multiplying its terms by an integer greater than one. In this way, numerous equivalent fractions may be obtained. Thus:

$$\frac{2}{3} = \frac{2 \times 8}{3 \times 8} = \frac{16}{24}, \text{ or } \frac{2}{3} = \frac{2 \times 30}{3 \times 30} = \frac{60}{90}$$

Two simple fractions can generally be changed to fractions equivalent to the given ones, and having a *common denominator*. For example, $\frac{2}{3}$ and $\frac{3}{4}$ are equivalent to $\frac{8}{12}$ and $\frac{9}{12}$. The fractions $\frac{8}{12}$ and $\frac{9}{12}$ may now be added to give $\frac{17}{12}$, since the 8 and the 9 both count the same thing, namely, twelfths.

Again, $\frac{5}{8}$ and $\frac{7}{12}$ are equivalent to $\frac{5 \times 3}{8 \times 3}$ and $\frac{7 \times 2}{12 \times 2}$, or $\frac{5}{8}$ and $\frac{7}{12}$ are equivalent to $\frac{15}{24}$ and $\frac{14}{24}$, respectively. The difference between $\frac{5}{8}$ and $\frac{7}{12}$ is thus found to be $\frac{1}{24}$, since we can subtract $\frac{14}{24}$ from $\frac{15}{24}$.

FRACTIONS

EXERCISE 1

Change the following fractions to equivalent fractions with higher terms or lower terms, as required. (Fill in the spaces.)

1. $\dfrac{3}{4} = \dfrac{-}{8}$, $\quad \dfrac{3}{4} = \dfrac{-}{12}$, $\quad \dfrac{3}{4} = \dfrac{-}{100}$, $\quad \dfrac{3}{4} = \dfrac{24}{-}$.
2. Change $\frac{1}{2}$ and $\frac{1}{3}$ both to 6ths.
3. Change $\frac{1}{3}$ and $\frac{3}{4}$ both to 12ths.
4. Change $\frac{3}{4}$ and $\frac{2}{5}$ both to 20ths.
5. Change $\frac{2}{3}$ and $\frac{3}{5}$ both to 15ths.
6. Change $\frac{1}{2}$ and $\frac{4}{5}$ both to 10ths.
7. Change $\frac{3}{4}$ and $\frac{5}{6}$ both to 12ths.
8. Change $\frac{3}{5}$ and $\frac{1}{6}$ both to 30ths.
9. Change $\frac{3}{5}$ and $\frac{3}{7}$ both to 35ths.
10. Change $\frac{3}{4}$ and $\frac{2}{7}$ both to 28ths.
11. Change $\frac{8}{40}$, $\frac{8}{10}$, $\frac{10}{25}$, $\frac{1}{1}$, $\frac{16}{10}$, $\frac{81}{45}$, and $\frac{50}{125}$ all to 5ths.
12. Change $\frac{2}{12}$, $\frac{3}{18}$, $\frac{30}{36}$, $\frac{1}{1}$, $\frac{66}{36}$, $\frac{40}{48}$, and $\frac{36}{54}$ all to 6ths.
13. Change $\frac{2}{3}$, $\frac{10}{18}$, $\frac{12}{27}$, $\frac{6}{54}$, $\frac{7}{3}$, $\frac{1}{1}$, and $\frac{3}{1}$ all to 9ths.
14. Change $\frac{1}{4}$, $\frac{2}{3}$, $\frac{5}{12}$, $\frac{1}{2}$, $\frac{3}{8}$, $\frac{9}{4}$, $\frac{14}{48}$, and $\frac{2}{1}$ to 24ths.
15. Change $\frac{1}{4}$, $\frac{5}{16}$, $\frac{3}{8}$, $\frac{11}{8}$, $\frac{1}{2}$, $\frac{3}{4}$, $\frac{10}{64}$, and $\frac{3}{1}$ to 32nds.
16. How many quarts are there in 1 and $\frac{3}{4}$ gallons?
17. How many pints are there in 3 quarts? How many pints are there in 5 gallons?
18. How many ounces are there in $1\frac{3}{4}$ pounds? (16 ounces equal 1 pound.)
19. Which is larger, two-thirds of 1 foot or three-fourths of 1 foot? How much is the difference?
20. Change $\frac{2}{5}$ and $\frac{3}{8}$ both to 40ths.
21. Change $\frac{5}{16}$ and $\frac{4}{9}$ both to 144ths.
22. Change $\frac{2}{3}$, $\frac{5}{8}$, $\frac{15}{48}$, $\frac{3}{16}$, 1, and 4 all to 48ths.
23. How many quarts are there in $3\frac{1}{4}$ gallons?
24. How many pints are there in 10 quarts? How many pints are there in 15 gallons?

25. How many ounces are there in $5\frac{3}{4}$ pounds?

26. Which is larger, three-fifths of 1 mile or four-sevenths of 1 mile? How much is the difference?

27. Change the following fractions to equivalent fractions with higher terms or lower terms, as required. (Fill in the spaces.)

$$\frac{5}{6} = \frac{}{30}$$

$$\frac{5}{6} = \frac{40}{}$$

$$\frac{3}{10} = \frac{}{100}$$

$$\frac{9}{25} = \frac{108}{}$$

28. Change the following fractions to equivalent fractions with higher terms or lower terms, as required. (Fill in the spaces.)

$$\frac{3}{8} = \frac{}{32}$$

$$\frac{3}{8} = \frac{18}{}$$

$$\frac{4}{9} = \frac{}{81}$$

$$\frac{9}{16} = \frac{45}{}$$

29. Change $\frac{3}{5}$ and $\frac{5}{16}$ both to 80ths.

30. How many gallons are there in 60 quarts?

If two fractions have the same denominator we may add them by adding the numerators. For instance:

$$\frac{1}{5} + \frac{2}{5} = \frac{3}{5}$$

If the fractions do not have the same denominator we must make an adjustment that will give them the same denominator. In the case of $\frac{1}{2} + \frac{3}{4}$, for example, we change the $\frac{1}{2}$ to $\frac{2}{4}$ by multiplying both numerator and denominator by 2. That is, $\frac{1}{2} \times \frac{2}{2} = \frac{2}{4}$. Then $\frac{1}{2} + \frac{3}{4}$ becomes $\frac{2}{4} + \frac{3}{4} = \frac{5}{4}$.

FRACTIONS

Another approach to be used in adding fractions follows: Consider now $\frac{2}{3} + \frac{5}{18} + \frac{9}{25}$. Find the *Lowest Common Denominator* (L.C.D.) of the three denominators. The lowest common denominator is the smallest quantity divisible by all the denominators under consideration. In order to proceed further, let us break each denominator into its prime factors. A *factor* of a number is any exact divisor of the number. Thus, the factors of 18 are 1, 2, 3, 6, 9, and 18. *Prime factors* are factors that may be divided only by themselves and 1. In the factors of 18, for example, 1, 2, and 3 are primes, since they may be divided only by themselves and 1. Six is not a prime, since it may be divided by itself, 1, 2, and 3. Nine is not a prime, since it may be divided by itself, 1, and 3.

Consider $\frac{3}{8} + \frac{5}{12}$. Let us write 8 in terms of its prime factors, $2 \cdot 2 \cdot 2$. We also shall write 12 in terms of its prime factors $2 \cdot 2 \cdot 3$. Thus:

$$\frac{3}{8} + \frac{5}{12} = \frac{3}{2 \cdot 2 \cdot 2} + \frac{5}{2 \cdot 2 \cdot 3}$$

In finding the lowest common denominator of the denominators 8 and 12, we must find a denominator that provides for all prime factors the greatest number of times each prime factor appears in each denominator. Now let us write the lowest common denominator of the denominators $2 \cdot 2 \cdot 2$ and $2 \cdot 2 \cdot 3$. This lowest common denominator is $2 \cdot 2 \cdot 2 \cdot 3$, rather than $2 \cdot 2 \cdot 2 \cdot 2 \cdot 3$. We do not list the 2 twos under the 5 because they are already provided for in the 3 twos which appear as prime factors of the denominator $2 \cdot 2 \cdot 2$. We list the 3 that appears under the 5; otherwise we would not make provision for it in the lowest common denominator.

Now:

$$\frac{3}{8} + \frac{5}{12} = \frac{3}{2 \cdot 2 \cdot 2} + \frac{5}{2 \cdot 2 \cdot 3} = \frac{3 \cdot 3}{2 \cdot 2 \cdot 2 \cdot 3} + \frac{5 \cdot 2}{2 \cdot 2 \cdot 2 \cdot 3}$$

$$= \frac{3 \cdot 3 + 5 \cdot 2}{2 \cdot 2 \cdot 2 \cdot 3} = \frac{9 + 10}{24} = \frac{19}{24}$$

Looking again at the example $\frac{2}{3} + \frac{5}{18} + \frac{9}{25}$, we express each denominator in terms of prime factors. Then:

$$\frac{2}{1 \cdot 3} + \frac{5}{2 \cdot 3 \cdot 3} + \frac{9}{5 \cdot 5} = ?$$

In finding the lowest common denominator of a series of denominators, we must write a denominator that provides for all factors the greatest number of times each factor appears in each denominator. Now let us write the lowest common denominator of the denominators $1 \cdot 3$, $2 \cdot 3 \cdot 3$, and $5 \cdot 5$. We will not write two threes to the right of the two because there is already one three in the set of numbers in the lowest common denominator.

Then:

$$\frac{2}{1 \cdot 3} + \frac{5}{2 \cdot 3 \cdot 3} + \frac{9}{5 \cdot 5} =$$

$$\frac{?}{1 \cdot 3 \cdot 2 \cdot 3 \cdot 5 \cdot 5} =$$

$$\frac{2(2 \cdot 3 \cdot 5 \cdot 5) + 5(5 \cdot 5) + 9(1 \cdot 3 \cdot 2 \cdot 3)}{1 \cdot 3 \cdot 2 \cdot 3 \cdot 5 \cdot 5} =$$

$$\frac{300 + 125 + 162}{450} = \frac{587}{450}$$

Let us go through the steps in the foregoing simplification. We find that the first denominator, $1 \cdot 3$, if it were to be made like the lowest common denominator, would need to be multiplied by $2 \cdot 3 \cdot 5 \cdot 5$. Therefore, we multiply the numerator by $2 \cdot 3 \cdot 5 \cdot 5$ in order to treat the numerator in the same way we treated the denominator. We see that the second denominator, $2 \cdot 3 \cdot 3$, if it were to be made like the lowest common denominator, would need to be multiplied by $5 \cdot 5$. Therefore, we multiply the numerator by $5 \cdot 5$. We observe that the third denominator, $5 \cdot 5$, if it were to be made like the L.C.D., would need to be multiplied by $3 \cdot 2 \cdot 3$. Since all denominators are alike, we may add the numerators and simplify as indicated in the foregoing illustration.

The fraction $\frac{5}{4}$ we have already referred to as an improper fraction. The expression $1\frac{1}{4}$ is a mixed number. A mixed number is the sum of an integer and a simple fraction. By integer we mean any of the numbers 1, 2, 3, etc. Notice that $1\frac{1}{4}$ means $1 + \frac{1}{4}$.

In subtracting fractions the operation with the denominators is identical with that of adding fractions. However, the numerators are subtracted rather than added.

Consider $\frac{5}{6} - \frac{2}{3} = ?$ The lowest common denominator of 6 and 3 is 6. Now the denominator 6 goes into the lowest common denominator 6, one time and 1 times 5 equals 5. The denominator 3 goes

FRACTIONS

into the lowest common denominator 6, two times and 2 times 2 equals 4. Now:

$$\frac{5}{6} - \frac{2}{3} = \frac{5}{6} - \frac{4}{6} = \frac{5-4}{6} = \frac{1}{6}$$

Another example is $\frac{2}{3} + \frac{3}{8} - \frac{3}{5} = ?$ The lowest common denominator for the 3, 8, and 5 is 120. Then:

$$\frac{2}{3} + \frac{3}{8} - \frac{3}{5} = \frac{80}{120} + \frac{45}{120} - \frac{72}{120} = \frac{80 + 45 - 72}{120}$$

$$= \frac{125 - 72}{120} = \frac{53}{120}$$

EXERCISE 2

Add and subtract as indicated.

1. $\dfrac{1}{3} + \dfrac{1}{3} =$
2. $\dfrac{1}{4} + \dfrac{1}{4} =$
3. $\dfrac{1}{6} + \dfrac{1}{6} =$
4. $\dfrac{3}{5} + \dfrac{2}{5} =$
5. $\dfrac{3}{7} + \dfrac{4}{7} =$
6. $\dfrac{6}{15} + \dfrac{9}{15} =$
7. $\dfrac{2}{3} - \dfrac{1}{3} =$
8. $\dfrac{7}{8} - \dfrac{3}{8} =$
9. $\dfrac{4}{5} - \dfrac{2}{5} =$
10. $\dfrac{2}{3} + \dfrac{3}{5} =$
11. $\dfrac{1}{2} + \dfrac{3}{4} =$
12. $\dfrac{5}{6} + \dfrac{3}{8} =$
13. $\dfrac{2}{7} + \dfrac{5}{8} =$
14. $\dfrac{3}{4} + \dfrac{5}{8} =$
15. $\dfrac{3}{4} - \dfrac{1}{2} =$
16. $\dfrac{7}{8} - \dfrac{1}{4} =$
17. $\dfrac{9}{16} - \dfrac{1}{4} =$
18. $\dfrac{11}{12} - \dfrac{3}{4} =$

19. $\dfrac{3}{7} + \dfrac{3}{4} + \dfrac{5}{7} =$

20. $\dfrac{1}{5} + \dfrac{2}{3} + \dfrac{3}{5} =$

21. $\dfrac{5}{16} + \dfrac{2}{3} + \dfrac{3}{5} =$

22. $\dfrac{5}{8} + \dfrac{2}{3} - \dfrac{3}{4} =$

23. $\dfrac{7}{16} + \dfrac{2}{5} - \dfrac{3}{8} =$

24. $\dfrac{3}{7} + \dfrac{5}{8} - \dfrac{1}{4} =$

25. $\dfrac{2}{9} + \dfrac{5}{16} + \dfrac{5}{8} + \dfrac{7}{9} =$

26. $\dfrac{5}{7} + \dfrac{4}{5} + \dfrac{7}{8} + \dfrac{2}{3} =$

27. $\dfrac{2}{5} - \dfrac{3}{7} + \dfrac{15}{45} - \dfrac{1}{4} =$

28. $\dfrac{5}{19} - \dfrac{4}{17} =$

29. $\dfrac{1}{5} + \dfrac{2}{3} + \dfrac{3}{15} + \dfrac{5}{6} + \dfrac{8}{9} =$

30. $\dfrac{3}{16} + \dfrac{1}{32} + \dfrac{7}{10} - \dfrac{3}{7} =$

31. $\dfrac{3}{4} + \dfrac{5}{9} - \dfrac{6}{17} - \dfrac{3}{19} =$

32. $\dfrac{4}{15} + \dfrac{2}{3} + \dfrac{3}{7} + \dfrac{8}{9} - \dfrac{3}{16} =$

33. The heirs to the Johnson estate receive shares of $\frac{1}{4}$, $\frac{1}{8}$, $\frac{2}{9}$, and $\frac{1}{3}$. The rest is left to charities. What fraction of the estate is left to charities? (Add the shares, and subtract their sum from 1.)

34. In sharing the annual bonus, four employees receive shares as follows: $\frac{1}{10}$, $\frac{2}{5}$, $\frac{1}{8}$, and $\frac{3}{20}$ respectively. What fraction will represent the share of the fifth employee? Assume that the five share the whole bonus.

35. A father dies and leaves his camels to three sons, Ralph, Reb, and Ruel. He leaves one-half of the animals to Ralph, one-third to Reb, and one-ninth to Ruel. At his death it is found that there are 17 camels. Since no one wants to divide a camel, the problem of fair distribution was given to Mahlon, the mathematician. Mahlon divided the camels in such a way that the number received by Ralph exceeded one-half of 17, the number received by Reb exceeded one-third of 17, and the number received by Ruel exceeded one-ninth of 17. How many did each receive?

36. How much larger is $\frac{9}{16}$ than $\frac{3}{8}$?

37. How much larger is $\frac{3}{4}$ than $\frac{5}{8}$?

38. One budget suggests $\frac{3}{20}$ of income for rent, $\frac{1}{4}$ for groceries, $\frac{1}{8}$ for meat, and $\frac{7}{50}$ for clothing. How much remains for other purposes?

39. Another budget suggests that $\frac{1}{5}$ of income be spent for taxes,

FRACTIONS   **41**

$\frac{1}{4}$ for rent, and $\frac{1}{20}$ for insurance. What part of the income is left for other uses?

40. A carpenter has drill bits with diameters of $\frac{7}{12}$, $\frac{11}{18}$, $\frac{1}{2}$, and $\frac{5}{9}$ inches. Arrange these fractions in order from largest to smallest.

41. Arrange the following fractions in order from largest to smallest: $\frac{5}{18}$, $\frac{3}{25}$, $\frac{4}{7}$, and $\frac{9}{16}$.

42. Find the total of $\frac{3}{8} + \frac{4}{9} - \frac{5}{36} + \frac{7}{18}$, and subtract it from the total of $\frac{5}{6} + \frac{7}{12} + \frac{17}{18} - \frac{7}{16}$.

2 | Multiplying proper fractions

In multiplying two proper fractions we multiply the two numerators to obtain a new numerator and multiply the two denominators to obtain a new denominator. The resulting fraction should be reduced to lowest terms. Thus:

$$\frac{3}{8} \times \frac{2}{15} = \frac{6}{120} = \frac{1}{20}$$

It is frequently preferred to simplify by division as follows:

$$\frac{\cancel{3}^{1}}{\cancel{8}_{4}} \times \frac{\cancel{2}^{1}}{\cancel{15}_{5}} = \frac{1}{20}$$

Here we have divided the 3 into itself and into the 15. The 3 goes into itself one time and into the 15 five times. The two goes into itself one time and into the 8 four times. Multiplying the two new numerators, we have $1 \times 1 = 1$. Multiplying the two new denominators, we have $4 \times 5 = 20$. The final result is $\frac{1}{20}$.

As another example, consider

$$\frac{2}{3} \times \frac{4}{5} \times \frac{25}{48} \times \frac{9}{14}$$

Let us factor each numerator and denominator into prime factors.

Now we have

$$\frac{2}{3} \times \frac{2 \cdot 2}{1 \cdot 5} \times \frac{5 \cdot 5}{2 \cdot 2 \cdot 2 \cdot 3} \times \frac{3 \cdot 3}{2 \cdot 7} =$$

$$\frac{\overset{1}{\cancel{2}} \times \overset{1}{\cancel{2}} \times \overset{1}{\cancel{2}} \times \overset{1}{\cancel{5}} \times 5 \times \overset{1}{\cancel{3}} \times \overset{1}{\cancel{3}}}{\underset{1}{\cancel{3}} \times \underset{1}{\cancel{5}} \times \underset{1}{\cancel{2}} \times \underset{1}{\cancel{2}} \times \underset{1}{\cancel{2}} \times 2 \times \underset{1}{\cancel{3}} \times 2 \times 7} = \frac{5}{28}$$

Here, the numerator, $2 \times 2 \times 2 \times 5 \times 5 \times 3 \times 3$, consists of all the factors of the numerators, and the denominator, $3 \times 5 \times 2 \times 2 \times 2 \times 2 \times 3 \times 2 \times 7$, consists of all the factors of the denominators. Reduction to lowest terms is simply the dividing out of all factors common to both numerator and denominator and multiplying the remaining numerator terms for a new numerator and the remaining denominator terms for a new denominator. Thus we obtain $\frac{5}{28}$.

For example, consider

$$\frac{3}{4} \times \frac{8}{15} \times \frac{25}{64} \times \frac{18}{55} =$$

$$\frac{3}{2 \cdot 2} \times \frac{2 \cdot 2 \cdot 2}{3 \cdot 5} \times \frac{5 \cdot 5}{2 \cdot 2 \cdot 2 \cdot 2 \cdot 2} \times \frac{2 \cdot 3 \cdot 3}{5 \cdot 11} =$$

$$\frac{\cancel{2} \cdot \cancel{2} \cdot \cancel{2} \cdot \cancel{2} \cdot \cancel{3} \cdot 3 \cdot 3 \cdot \cancel{5} \cdot \cancel{5}}{\cancel{2} \cdot \cancel{2} \cdot \cancel{2} \cdot \cancel{2} \cdot 2 \cdot 2 \cdot 2 \cdot 2 \cdot \cancel{3} \cdot \cancel{5} \cdot \cancel{5} \cdot 11} = \frac{9}{176}$$

Dividing the numerator by 2 and the denominator by 2 until all of the 2's are removed from either the numerator or the denominator, we have

$$\frac{\cancel{3} \cdot 3 \cdot 3 \cdot 5 \cdot 5}{2 \cdot 2 \cdot 2 \cdot 2 \cdot \cancel{3} \cdot 5 \cdot 5 \cdot 11}$$

Dividing the numerator by 3 and the denominator by 3 until all of the 3's are removed from either the numerator or the denominator, we have

$$\frac{3 \cdot 3 \cdot \cancel{5} \cdot \cancel{5}}{2 \cdot 2 \cdot 2 \cdot 2 \cdot \cancel{5} \cdot \cancel{5} \cdot 11}$$

Dividing the numerator by 5 and the denominator by 5 until all of the 5's are removed from the numerator and the denominator, we have

$$\frac{3 \cdot 3}{2 \cdot 2 \cdot 2 \cdot 2 \cdot 11} = \frac{9}{176}$$

FRACTIONS _____43

It should be noticed that in dividing 2 into 2, 3 into 3, 5 into 5, etc. we obtain 1 each time. However, 1 multiplied by any number gives that number. Consequently, the 1's are not indicated.

EXERCISE 3

Multiply and reduce the answer to lowest terms.

1. $\dfrac{2}{3} \times \dfrac{3}{5} =$ 2. $\dfrac{3}{4} \times \dfrac{4}{9} =$

3. $\dfrac{1}{2} \times \dfrac{4}{7} =$ 4. $\dfrac{3}{5} \times \dfrac{2}{3} =$

5. $\dfrac{1}{2} \times \dfrac{2}{3} =$ 6. $\dfrac{5}{6} \times \dfrac{3}{7} =$

7. $\dfrac{4}{5} \times \dfrac{3}{8} =$ 8. $\dfrac{3}{10} \times \dfrac{5}{9} =$

9. $\dfrac{2}{5} \times \dfrac{5}{16} =$ 10. $\dfrac{5}{8} \times \dfrac{16}{45} =$

11. $\dfrac{4}{21} \times \dfrac{3}{8} =$ 12. $\dfrac{5}{6} \times \dfrac{3}{5} =$

13. $\dfrac{4}{5} \times \dfrac{5}{18} =$ 14. $\dfrac{5}{6} \times \dfrac{3}{25} =$

15. $\dfrac{7}{30} \times \dfrac{4}{49} =$ 16. $\dfrac{6}{25} \times \dfrac{5}{8} =$

17. $\dfrac{5}{68} \times \dfrac{6}{35} =$ 18. $\dfrac{15}{64} \times \dfrac{8}{45} =$

19. $\dfrac{23}{45} \times \dfrac{5}{46} =$ 20. $\dfrac{95}{162} \times \dfrac{81}{130} =$

21. $\dfrac{64}{175} \times \dfrac{55}{146} =$ 22. $\dfrac{8}{17} \times \dfrac{51}{56} =$

23. $\dfrac{2}{3} \times \dfrac{5}{12} \times \dfrac{5}{8} =$ 24. $\dfrac{5}{12} \times \dfrac{3}{10} \times \dfrac{4}{15} =$

25. $\dfrac{2}{5} \times \dfrac{1}{4} \times \dfrac{5}{8} =$ 26. $\dfrac{1}{4} \times \dfrac{8}{15} \times \dfrac{3}{4} =$

27. $\dfrac{6}{25} \times \dfrac{5}{8} \times \dfrac{3}{4} =$

28. $\dfrac{9}{44} \times \dfrac{6}{25} \times \dfrac{20}{81} =$

29. $\dfrac{3}{40} \times \dfrac{20}{99} \times \dfrac{40}{53} =$

30. $\dfrac{9}{50} \times \dfrac{25}{36} \times \dfrac{64}{91} =$

31. $\dfrac{12}{25} \times \dfrac{15}{84} \times \dfrac{56}{63} \times \dfrac{69}{144} =$

32. $\dfrac{47}{66} \times \dfrac{16}{45} \times \dfrac{92}{141} \times \dfrac{90}{164} =$

33. $\dfrac{14}{15} \times \dfrac{19}{24} \times \dfrac{45}{57} \times \dfrac{3}{56} =$

34. $\dfrac{65}{256} \times \dfrac{20}{47} \times \dfrac{64}{195} \times \dfrac{188}{260} =$

35. $\dfrac{6}{15} \times \dfrac{25}{48} \times \dfrac{56}{135} \times \dfrac{91}{100} \times \dfrac{20}{49} =$

36. $\dfrac{14}{27} \times \dfrac{81}{70} \times \dfrac{625}{900} \times \dfrac{256}{1000} =$

*37. $\dfrac{27}{52} \times \dfrac{830}{2433} \times \dfrac{29}{83} \times \dfrac{99}{594} =$

38. What is the cost of $4\tfrac{3}{4}$ yards of cloth at $1.82 per yard?

39. What is the value of 45 shares of stock at $78\tfrac{1}{8}$ per share?

40. What is the area of a room $15\tfrac{3}{4}$ feet long by $20\tfrac{2}{3}$ feet wide? Partial solution:

$$(15\tfrac{3}{4}) \times (20\tfrac{2}{3}) =$$

$$15 \times 20 + \dfrac{3}{4} \times 20 + \dfrac{2}{3} \times 15 + \dfrac{3}{4} \times \dfrac{2}{3} =$$

$$300 + 15 + 10 + \dfrac{1}{2} = \ ?$$

41. What is the perimeter of the room in Problem 40?

42. If a cubic foot of space holds $7\tfrac{1}{2}$ gallons, how many gallons will a tank hold if its volume is 90 cubic feet?

43. How many ounces are there in $31\tfrac{1}{2}$ pounds of glue?

44. What are the area and the perimeter of a room $18\tfrac{3}{4}$ by $12\tfrac{1}{2}$ feet?

45. Find the total value of the following lots of shares: 46 shares at $22\tfrac{1}{4}$, 58 shares at $17\tfrac{1}{8}$, 50 shares at $72\tfrac{3}{4}$, and 14 shares at $65\tfrac{1}{2}$.

46. What is the total weight of 72 candy bars each containing $1\tfrac{3}{8}$ ounces of candy?

47. What is the volume of a box $9\tfrac{3}{4}$ inches by $3\tfrac{1}{2}$ inches by $5\tfrac{1}{4}$ inches? (Volume = length × width × depth.)

FRACTIONS ───────────────────────────────── **45**

48. If 1 cubic foot of water weighs 62½ pounds, find the weight of a vat of water containing 48 cubic feet. (Ignore the weight of the container.)

49. Al in five hours a sum can count,
   Which Sal can in eleven;
   How much more then is the amount
   They both can count in seven?

50. An auditorium seats 2400 persons. At a recent entertainment one usher guessed it was three-fourths full, another that it was two-thirds full. The ticket office reported 1900 sales. Which usher (first or second) made the better guess?

51. If, on the plan of a house that the Springers are studying, ⅛ inch represents 1 foot, what are the dimensions of a room that measures 1¾ inches by 2½ inches on the plan?

52. If potatoes are four-fifths water, what is the weight of the water in 8 bushels? (Assume that a bushel of potatoes weighs 60 pounds.)

53. Alice bought a sheet of five-cent postage stamps (100) and used three-fifths of them the first week. How many did she have left?

54. What is the total weight of 96 candy bars each containing 1⅜ ounces of candy?

3 | Dividing proper fractions

In dividing simple fractions we may think of the process of division as *measurement* or *partition*, or of *changing to an equivalent fraction* with simpler terms. In each case we may change the problem from one of division to one of multiplication.

Consider the following examples:

A. ⅜ ÷ 2 can mean: find ½ of ⅜. Therefore we have

$$\frac{3}{8} \div 2 = \frac{1}{2} \times \frac{3}{8} = \frac{3}{16}$$

Notice that $\frac{3}{16}$ is one of two equal parts of $\frac{3}{8}$. This is the idea of *partition*.

B. 3 ÷ ¼ can mean find how many times ¼ is contained in 3. Since there are 4 quarters in one unit, 3 units contain 3 × 4 quarters, or 12 quarters. This is the idea of *measurement:* $\frac{1}{4}$ inch is contained in 3 inches, 12 times.

C. $\frac{3}{4} \div \frac{2}{3}$ is the same as the fraction $\frac{\frac{3}{4}}{\frac{2}{3}}$. Changing to higher terms we have

$$\frac{\frac{3}{4}}{\frac{2}{3}} = \frac{\frac{3}{4} \times 12}{\frac{2}{3} \times 12} = \frac{9}{8} = 1\frac{1}{8}$$

D. Again $\frac{3}{4} \div \frac{2}{3}$ is the same as $\frac{\frac{3}{4}}{\frac{2}{3}}$, and to change terms we write

$$\frac{\frac{3}{4}}{\frac{2}{3}} = \frac{\frac{3}{4} \times \frac{3}{2}}{\frac{2}{3} \times \frac{3}{2}} = \frac{\frac{3}{4} \times \frac{3}{2}}{1} = \frac{3}{4} \times \frac{3}{2} = \frac{9}{8} \text{ or } 1\frac{1}{8}$$

Notice that we have changed the problem in division, $\frac{3}{4} \div \frac{2}{3}$, to an equivalent problem in multiplication, $\frac{3}{4} \times \frac{3}{2}$. This procedure is always possible with simple fractions. Thus:

$$\frac{a}{b} \div \frac{c}{d} = \frac{\frac{a}{b}}{\frac{c}{d}} = \frac{\frac{a}{b} \times \frac{d}{c}}{\frac{c}{d} \times \frac{d}{c}} = \frac{\frac{a}{b} \times \frac{d}{c}}{\frac{c}{d} \times \frac{d}{c}} = \frac{\frac{a}{b} \times \frac{d}{c}}{1}$$

Therefore:

$$\frac{a}{b} \div \frac{c}{d} = \frac{a}{b} \times \frac{d}{c}$$

We speak of this procedure in division as inverting the divisor.

In dividing two proper fractions we invert the divisor and then multiply the two fractions.

For example:

$$\frac{3}{4} \div \frac{3}{8} = \frac{\cancel{3}^1}{\cancel{4}_1} \times \frac{\cancel{8}^2}{\cancel{3}_1} = 2$$

Here we have $\frac{3}{8}$ as the divisor. Upon inverting $\frac{3}{8}$, we obtain $\frac{8}{3}$.

The division process might also be expressed as

$$\frac{\frac{5}{6}}{\frac{25}{126}} = \frac{\cancel{5}^1}{\cancel{6}_1} \times \frac{\cancel{126}^{21}}{\cancel{25}_5} = \frac{21}{5} = 4\frac{1}{5}$$

FRACTIONS _____ 47

In a series of divisions of fractions, such as $\frac{3}{4} \div \frac{1}{2} \div \frac{2}{3}$, we have

$$\frac{3}{4} \times \frac{\cancel{2}}{1}^{1} \times \frac{3}{\cancel{2}_{1}} = \frac{9}{4} \quad \text{or} \quad 2\frac{1}{4}$$

EXERCISE 4

Divide and reduce the answer to lowest terms.

1. $\dfrac{1}{3} \div \dfrac{2}{3} =$ 2. $\dfrac{5}{9} \div \dfrac{5}{6} =$

3. $\dfrac{3}{16} \div \dfrac{9}{32} =$ 4. $\dfrac{4}{5} \div \dfrac{8}{15} =$

5. $\dfrac{2}{5} \div \dfrac{12}{25} =$ 6. $\dfrac{5}{12} \div \dfrac{15}{4} =$

7. $\dfrac{2}{3} \div 6 =$ 8. $\dfrac{4}{5} \div 4 =$

9. $\dfrac{3}{5} \div 12 =$ 10. $\dfrac{3}{8} \div 9 =$

11. $\dfrac{3}{4} \div 15 =$ 12. $\dfrac{5}{8} \div 10 =$

13. $\dfrac{3}{5} \div \dfrac{12}{25} =$ 14. $\dfrac{5}{11} \div \dfrac{10}{33} =$

15. $\dfrac{2}{3} \div \dfrac{5}{3} =$ 16. $\dfrac{4}{5} \div \dfrac{16}{25} =$

17. $\dfrac{5}{6} \div \dfrac{35}{42} =$ 18. $\dfrac{2}{5} \div \dfrac{18}{25} =$

19. $\dfrac{14}{27} \div \dfrac{4}{3} =$ 20. $\dfrac{15}{22} \div \dfrac{3}{2} =$

21. $\dfrac{3}{8} \div \dfrac{15}{16} =$ 22. $\dfrac{5}{16} \div \dfrac{25}{64} =$

23. $\dfrac{3}{4} \div \dfrac{27}{86} =$ 24. $\dfrac{4}{9} \div \dfrac{16}{81} =$

25. $\dfrac{7}{8} \div \dfrac{21}{24} =$

26. $\dfrac{5}{8} \div \dfrac{45}{32} =$

27. $\dfrac{17}{25} \div \dfrac{68}{15} =$

28. $\dfrac{\frac{3}{4}}{\frac{15}{16}} =$

29. $\dfrac{\frac{2}{3}}{\frac{16}{9}} =$

30. $\dfrac{\frac{4}{9}}{\frac{48}{27}} =$

31. $\dfrac{\frac{5}{8}}{\frac{25}{16}} =$

32. $\dfrac{\frac{7}{15}}{\frac{56}{75}} =$

33. $\dfrac{\frac{11}{20}}{\frac{99}{80}} =$

34. $\dfrac{\frac{15}{64}}{\frac{105}{192}} =$

*35. $\dfrac{3}{4} \div \dfrac{5}{4} \div \dfrac{21}{35} =$

*36. $\dfrac{7}{8} \div \dfrac{35}{24} \div \dfrac{42}{125} =$

*37. $\dfrac{5}{6} \div \dfrac{135}{156} \div \dfrac{65}{108} =$

*38. $\dfrac{\frac{5}{8} \div \frac{15}{16}}{\frac{3}{7} \div \frac{27}{98}} =$

*39. $\dfrac{\frac{4}{5} \div \frac{84}{95}}{\frac{3}{4} \div \frac{75}{88}} =$

*40. $\dfrac{128}{265} \div \dfrac{512}{1325} \div \dfrac{95}{16} =$

*41. $\dfrac{360}{729} \div \dfrac{126}{154} \div \dfrac{261}{342} =$

*42. $\dfrac{5}{18} \div \dfrac{4}{15} \div \dfrac{6}{85} \div \dfrac{25}{48} =$

*43. $\dfrac{3}{40} \div \dfrac{9}{20} \div \dfrac{55}{729} \div \dfrac{81}{192} =$

*44. $\dfrac{8}{25} \div \dfrac{96}{625} \div \dfrac{160}{192} \div \dfrac{22}{54} =$

45. How many pieces $1\frac{3}{4}$ inches in length can be cut from a bar 4 feet long if $\frac{1}{8}$ inch is allowed for each cut?

46. If a car uses $\frac{3}{32}$ gallon of gasoline per mile, how many miles can it travel on 54 gallons of gasoline?

47. If a car uses $\frac{3}{53}$ gallon of gasoline per mile, how many miles does it get to the gallon?

48. A piece of wood $\frac{25}{16}$ inch thick is made by gluing together strips each $\frac{5}{32}$ inch thick. How many strips are there in this piece?

49. A tank can be three-fifths filled in one hour. How long will it take to fill the tank?

50. A man can do three-eighths of a job in one day. How long will it take him to do the whole job?

51. A piece of wood $\frac{9}{8}$ inch thick is made by gluing together strips each $\frac{27}{72}$ inch thick. How many strips are there in this piece?

FRACTIONS

52. Abe can mow four-fifths of a lawn in three-fifths of a day. How long will it take him to mow the whole lawn?

53. A student can do one-fourth of an assignment in $\frac{3}{4}$ hour. How long will it take him to do the whole assignment?

54. How many pieces $\frac{7}{16}$ inch in length can be cut from a bar $3\frac{1}{2}$ feet in length if $\frac{1}{8}$ inch is allowed for each cut?

4 | Adding and subtracting improper fractions and mixed numbers

As we have already explained, an improper fraction is a fraction whose denominator is smaller than its numerator. Thus $\frac{9}{5}$ is an improper fraction.

In adding and subtracting improper fractions we proceed as we did in the case of proper fractions. For example, $\frac{8}{3} + \frac{14}{3} = ?$ Here the denominators are alike to begin with, and we add the two numerators 8 and 14 to obtain 22. Thus the answer is $\frac{22}{3}$, as an improper fraction, or $7\frac{1}{3}$ as a mixed number.

Now consider $\frac{15}{4} + \frac{13}{5} = ?$ Here the denominators are different. We find the lowest common denominator, which is 20. Now, as indicated before in adding fractions, the denominator 4 goes into the denominator 20, five times. We take five times 15 to obtain 75. The denominator 5 goes into 20, four times. We take four times 13 which is 52. Written in much more convenient form we have

$$\frac{15}{4} + \frac{13}{5} = \frac{75 + 52}{20} = \frac{127}{20} = 6\frac{7}{20}$$

As an example of subtracting improper fractions we have $\frac{5}{3} - \frac{6}{5} = ?$ The lowest common denominator of 3 and 5 is 15. The denominator 3 goes into the common denominator 15 five times. Five times the numerator 5 is 25. The denominator 5 goes into the common denominator 15 three times. Three times the numerator 6 is 18. Thus we have in a more convenient form:

$$\frac{25 - 18}{15} = \frac{7}{15}$$

In adding mixed numbers we have three types of approaches. In the first approach we arrange the numbers in a vertical line. In the second we arrange the numbers in a horizontal line. In the third we change the mixed numbers to improper fractions and proceed as in the illustrative examples just given.

In adding $65\frac{3}{4}$ we have the numbers arranged in a vertical line.
$42\frac{7}{8}$

In working this example consider first the proper fractions $\frac{3}{4}$ and $\frac{7}{8}$. Find the common denominator of 4 and 8, which is 8. Dividing the denominator 4 into the common denominator 8 we obtain 2. This number we multiply by the numerator 3, which gives 6. Thus the fraction $\frac{3}{4}$ is changed to an equivalent fraction or $\frac{6}{8}$. Now the problem of adding $65\frac{3}{4}$ becomes $65\frac{6}{8}$. Adding the $\frac{6}{8}$ and $\frac{7}{8}$ we have
$42\frac{7}{8}$ $42\frac{7}{8}$
$\overline{}$ $\overline{107\frac{13}{8}}$

$\frac{13}{8}$. Adding the 65 and 42, we obtain 107. The improper fraction $\frac{13}{8}$ simplifies into $1\frac{5}{8}$. Adding the 1 to 107 we obtain 108, and we write the final answer as $108\frac{5}{8}$.

In combining $65\frac{3}{4} + 42\frac{7}{8}$, we have the numbers arranged in a horizontal line. We may begin by taking the proper fractions $\frac{3}{4}$ and $\frac{7}{8}$ and finding the lowest common denominator, 8. Now, as in the preceding example, we convert $\frac{3}{4}$ to its equivalent fraction $\frac{6}{8}$. When we add the $\frac{6}{8}$ and the $\frac{7}{8}$, the result is $\frac{13}{8}$ or $1\frac{5}{8}$. Adding the 65 and the 42 we have 107. Then combining the 1 in the $1\frac{5}{8}$ with the 107, we obtain
$$65\frac{3}{4} + 42\frac{7}{8} = 108\frac{5}{8}$$

Finally, in adding $65\frac{3}{4}$ and $42\frac{7}{8}$, we may change each of the mixed numbers to improper fractions and combine, as in the section on improper fractions. Let us examine how to change the mixed number $65\frac{3}{4}$ to its equivalent improper fraction. $65\frac{3}{4}$ means 65 plus $\frac{3}{4}$. We must therefore change 65 to fourths in order to add it to $\frac{3}{4}$. Sixty-five equals 260 fourths. Now:

$$\frac{260}{4} + \frac{3}{4} = \frac{263}{4}$$

Thus:
$$65\frac{3}{4} + 42\frac{7}{8} = \frac{263}{4} + \frac{343}{8} = \frac{526 + 343}{8} = \frac{869}{8} = 108\frac{5}{8}$$

In subtracting mixed numbers we may follow any one of the three types of approach previously outlined for the addition of

FRACTIONS ─────────────────────────────────── 51

mixed numbers. In the first approach we arrange the numbers in a vertical line. In the second we arrange the numbers in a horizontal line. In the third we change the mixed numbers to improper fractions and then subtract. For example, as an illustration of arranging the numbers in a vertical line, subtract $83\frac{2}{7}$ from $245\frac{5}{6}$.

$$\begin{array}{c} 245\frac{5}{6} \\ 83\frac{2}{7} \end{array} \quad \text{becomes} \quad \begin{array}{c} 245\frac{35}{42} \\ 83\frac{12}{42} \\ \hline 162\frac{23}{42} \end{array}$$

The arrangement of numbers in a horizontal line is illustrated by

$$92\frac{3}{4} - 61\frac{5}{8} = 92\frac{6}{8} - 61\frac{5}{8} = 31\frac{1}{8}$$

An example of changing mixed numbers to improper fractions and then subtracting is

$$154\frac{3}{5} - 27\frac{3}{4} = \frac{773}{5} - \frac{111}{4} = \frac{3092 - 555}{20} = \frac{2537}{20} = 126\frac{17}{20}$$

Another example of subtracting mixed numbers is as follows. Subtract:

$$\begin{array}{c} 249\frac{3}{5} \\ 160\frac{5}{8} \end{array}$$

$$\begin{array}{c} 249\frac{3}{5} \\ 160\frac{5}{8} \end{array} \quad \text{becomes} \quad \begin{array}{c} 249\frac{24}{40} \\ 160\frac{25}{40} \end{array} \quad \text{or} \quad \begin{array}{c} 248\frac{64}{40} \\ 160\frac{25}{40} \\ \hline 88\frac{39}{40} \end{array}$$

EXERCISE 5

Add and subtract as indicated.

1. $3\frac{1}{2} + 7\frac{3}{4} =$
2. $9\frac{2}{5} + 6\frac{1}{2} =$
3. $12\frac{3}{8} + 17\frac{5}{16} =$
4. $16\frac{5}{14} + 15\frac{4}{7} =$
5. $23\frac{4}{5} + 19\frac{7}{15} =$
6. $72\frac{5}{6} + 18\frac{7}{18} =$
7. $18\frac{3}{5} + 19\frac{5}{6} =$
8. $37\frac{5}{8} + 49\frac{3}{7} =$
9. $69\frac{1}{2} - 42\frac{3}{4} =$
10. $57\frac{5}{8} - 14\frac{1}{4} =$
11. $86\frac{3}{5} - 49\frac{5}{6} =$
12. $115\frac{7}{16} - 98\frac{5}{9} =$

13. Add: $460\frac{13}{16}$
$138\frac{3}{4}$

14. Add: $392\frac{3}{8}$
$67\frac{5}{6}$

15. Add: $715\frac{3}{8}$
$263\frac{2}{3}$

16. Add: $842\frac{3}{5}$
$116\frac{4}{7}$

17. Subtract: $584\frac{3}{5}$
$229\frac{5}{8}$

18. Subtract: $795\frac{2}{3}$
$368\frac{3}{4}$

19. Subtract: $6184\frac{4}{15}$
$3927\frac{3}{7}$

20. Subtract: $9325\frac{7}{18}$
$643\frac{3}{4}$

21. $6\frac{3}{4} - 2\frac{5}{8} =$

22. $9\frac{5}{16} - 3\frac{7}{8} =$

23. $21\frac{5}{16} - 11\frac{3}{5} =$

24. $59\frac{5}{7} - 33\frac{4}{9} =$

25. $16\frac{5}{17} + 38\frac{3}{4} - 15\frac{3}{7} =$

26. $94\frac{3}{4} - 16\frac{2}{3} - 40\frac{3}{8} =$

27. $83\frac{9}{17} - 42\frac{3}{4} - 11\frac{1}{2} =$

28. $142\frac{3}{7} + 19\frac{4}{9} - 17\frac{3}{8} =$

29. $\dfrac{7}{4} + \dfrac{8}{5} =$

30. $\dfrac{11}{3} + \dfrac{9}{4} =$

31. $\dfrac{19}{5} + \dfrac{42}{11} + \dfrac{74}{9} =$

32. $\dfrac{65}{11} - \dfrac{17}{13} - \dfrac{9}{2} =$

33. $\dfrac{23}{3} + \dfrac{17}{5} - \dfrac{43}{17} =$

34. $\dfrac{31}{9} + \dfrac{47}{4} - \dfrac{13}{5} =$

35. $\dfrac{79}{14} - \dfrac{5}{3} - \dfrac{7}{4} =$

Add:

36. $\dfrac{11}{3}$
$\dfrac{7}{2}$

37. $\dfrac{42}{5}$
$\dfrac{7}{3}$
$\dfrac{9}{4}$

38. $\dfrac{164}{7}$
$\dfrac{15}{17}$
$\dfrac{3}{2}$

FRACTIONS

39. $\dfrac{17}{8}$
$\dfrac{23}{14}$
$\dfrac{73}{3}$
——

40. $\dfrac{163}{4}$
$\dfrac{523}{18}$
$\dfrac{625}{14}$
——

Subtract in each of the following problems.

41. $\dfrac{9}{5}$
$\dfrac{5}{3}$
——

42. $\dfrac{11}{3}$
$\dfrac{6}{5}$
——

43. $\dfrac{121}{49}$
$\dfrac{212}{147}$
——

44. $\dfrac{75}{14}$
$\dfrac{365}{196}$
——

*45. $\dfrac{85}{14}$
$\dfrac{365}{204}$
——

Combine as indicated:

46. $94\frac{3}{8} + \dfrac{160}{3} - 12\frac{4}{9} =$

47. $\dfrac{85}{7} - 4\frac{2}{5} - \dfrac{35}{18} =$

48. A piece of cloth $18\frac{3}{8}$ yards long shrank $1\frac{3}{8}$ yards in bleaching. How long was the cloth after bleaching?

49. The total length of two pieces of molding must be $16\frac{3}{4}$ feet. If one piece is $9\frac{2}{3}$ feet long, what is the proper length for the second piece?

50. A rope 36 feet long is desired. Two pieces, $8\frac{3}{8}$ feet and $12\frac{1}{4}$ feet long, are on hand. How long a piece must be bought, assuming that the pieces are not joined to form a closed circle and that each joining requires $\frac{5}{6}$ of a foot?

51. James is building a table top in his shop from 5 pieces of board. One board is $5\frac{1}{4}$ inches wide, the second is $5\frac{3}{8}$ inches wide, the third is $7\frac{7}{8}$ inches wide, and the fourth is $6\frac{3}{4}$ inches wide. How wide should the fifth board be in order for the top to be $31\frac{1}{2}$ inches wide?

*52. Rabbits in an experimental project are fed the following amounts on each of 7 days: $14\frac{3}{4}$ pounds, $13\frac{1}{2}$ pounds, $12\frac{4}{5}$ pounds,

15$\frac{7}{8}$ pounds, 14$\frac{1}{4}$ pounds, 10$\frac{3}{8}$ pounds, and 11$\frac{2}{3}$ pounds. If 100 pounds of feed were on hand to begin with, and $\frac{7}{8}$ of a pound were wasted during the 7-day period, how much feed was still on hand at the end of the 7 days?

53. A piece of goods 28$\frac{1}{2}$ yards long shrinks 2$\frac{1}{4}$ yards while being bleached. How long a piece remains after the bleaching process?

*54. Mr. Gibson has the following hours on his time card for a 10-day period: 5$\frac{1}{4}$, 8$\frac{1}{5}$, 7$\frac{3}{5}$, 9$\frac{4}{15}$, 11$\frac{3}{4}$, 10$\frac{7}{15}$, 12$\frac{5}{6}$, 9$\frac{1}{2}$, 10$\frac{1}{6}$, and 13$\frac{4}{5}$ hours. He is to get overtime for all hours in excess of 80. Does he have any overtime coming to him? If so, how many hours?

55. Baxter works the following hours on a part-time job for 5 days: 3$\frac{4}{5}$, 5$\frac{2}{3}$, 4$\frac{3}{5}$, 2$\frac{4}{15}$, and 5$\frac{1}{3}$. Find the total number of hours he works.

56. At Dr. Zimmerman's Veterinary Clinic the following amounts of dog feed are used each month for a year: 48$\frac{5}{8}$ pounds, 56$\frac{7}{16}$ pounds, 58$\frac{3}{4}$ pounds, 76$\frac{1}{2}$ pounds, 68$\frac{3}{4}$ pounds, 55$\frac{5}{8}$ pounds, 48$\frac{1}{8}$ pounds, 72$\frac{3}{8}$ pounds, 83$\frac{5}{16}$ pounds, 75$\frac{1}{4}$ pounds, 81$\frac{7}{8}$ pounds, and 75$\frac{1}{4}$ pounds. How many 50-pound sacks of dog food should be purchased to provide for the clinic during this 12-month period?

57. If 85$\frac{3}{4}$ feet of fencing are on hand and 146$\frac{2}{3}$ feet in frontage are to be fenced, how many feet must be purchased, allowing $\frac{5}{12}$ foot for overlapping of the 2 pieces?

5 | Multiplying improper fractions and mixed numbers

In multiplying improper fractions we may follow one of several different approaches. (1) We may multiply the numerators for a new numerator and multiply the denominators for a new denominator. The resulting fraction should always be reduced to lowest terms. For example: $\frac{9}{4} \times \frac{32}{27} = \frac{288}{108} = \frac{8}{3} = 2\frac{2}{3}$. Here we multiplied 32 and 9 to obtain 288 and multiplied 4 by 27 to obtain 108. (2) We can multiply improper fractions in the following way: Break up each numerator and each denominator into prime factors. Thus:

$$\frac{9}{4} \times \frac{32}{27} = \frac{3 \cdot 3}{2 \cdot 2} \times \frac{2 \cdot 2 \cdot 2 \cdot 2 \cdot 2}{3 \cdot 3 \cdot 3} = \frac{3 \cdot 3 \cdot 2 \cdot 2 \cdot 2 \cdot 2 \cdot 2}{2 \cdot 2 \cdot 3 \cdot 3 \cdot 3}$$

The fraction is reduced to lowest terms by dividing out all factors common to both numerator and denominator and multiplying the

FRACTIONS _____ 55

remaining numerator terms for a new numerator and the remaining denominator terms for a new denominator. Now:

$$\frac{\overset{1}{\cancel{3}}\cdot\overset{1}{\cancel{3}}\cdot\overset{1}{\cancel{2}}\cdot\overset{1}{\cancel{2}}\cdot 2 \cdot 2 \cdot 2}{\underset{1}{\cancel{2}}\cdot\underset{1}{\cancel{2}}\cdot\underset{1}{\cancel{3}}\cdot\underset{1}{\cancel{3}}\cdot 3} = \frac{8}{3} \quad \text{or} \quad 2\frac{2}{3}$$

(3) We may change the improper fractions to mixed numbers and proceed to multiply as follows: In the example $\frac{9}{4} \times \frac{32}{27} = ?$ change the $\frac{9}{4}$ to $2\frac{1}{4}$ and the $\frac{32}{27}$ to $1\frac{5}{27}$. Then $2\frac{1}{4} \times 1\frac{5}{27} = ?$ Multiply 2×1 to obtain 2. Multiply $\frac{1}{4} \times \frac{5}{27}$ to obtain $\frac{5}{108}$. Multiply $2 \times \frac{5}{27}$ to obtain $\frac{10}{27}$. Multiply $1 \times \frac{1}{4}$ to obtain $\frac{1}{4}$. Now add 2, $\frac{5}{108}$, $\frac{10}{27}$, and $\frac{1}{4}$, or

$$2 + \frac{5}{108} + \frac{10}{27} + \frac{1}{4} = 2 + \frac{5 + 40 + 27}{108} = 2 + \frac{72}{108} = 2\frac{2}{3}$$

Conversely, mixed numbers may be changed to improper fractions before multiplying, and the operation goes forward as already illustrated. Thus:

$$6\frac{3}{4} \times 1\frac{13}{15} = \frac{\overset{9}{\cancel{27}}}{\underset{1}{\cancel{4}}} \times \frac{\overset{7}{\cancel{28}}}{\underset{5}{\cancel{15}}} = \frac{63}{5} = 12\frac{3}{5}$$

EXERCISE 6

Multiply:

1. $\dfrac{5}{3} \times \dfrac{9}{5} =$
2. $\dfrac{7}{4} \times \dfrac{16}{3} =$
3. $\dfrac{11}{5} \times \dfrac{9}{4} =$
4. $\dfrac{13}{6} \times \dfrac{11}{9} =$
5. $\dfrac{21}{5} \times \dfrac{15}{7} =$
6. $\dfrac{24}{13} \times \dfrac{65}{6} =$
7. $\dfrac{32}{9} \times \dfrac{21}{16} =$
8. $\dfrac{55}{4} \times \dfrac{24}{11} =$
9. $\dfrac{112}{9} \times \dfrac{45}{16} =$
10. $\dfrac{225}{14} \times \dfrac{56}{25} =$

11. $3\frac{1}{3} \times 1\frac{1}{5} =$
12. $6\frac{2}{3} \times 2\frac{2}{5} =$
13. $4\frac{3}{8} \times 3\frac{3}{5} =$
14. $1\frac{1}{4} \times 4\frac{2}{5} =$
15. $3\frac{1}{3} \times 5\frac{7}{16} =$
16. $4\frac{7}{8} \times 6\frac{2}{3} =$
17. $5\frac{2}{5} \times 2\frac{2}{3} =$
18. $1\frac{2}{5} \times 3\frac{4}{7} =$
19. $2\frac{1}{2} \times 3\frac{1}{5} =$
20. $4\frac{1}{6} \times 3\frac{3}{5} =$
21. $5\frac{3}{4} \times 8\frac{1}{2} =$
22. $9\frac{2}{3} \times 7\frac{1}{4} =$
23. $12\frac{3}{5} \times 9\frac{3}{8} =$
24. $16\frac{5}{8} \times 6\frac{7}{16} =$
25. $24\frac{3}{7} \times 13\frac{1}{2} =$
26. $19\frac{4}{5} \times 24\frac{5}{16} =$
27. $35\frac{3}{16} \times 28\frac{4}{5} =$
28. $\dfrac{9}{4} \times \dfrac{16}{3} \times \dfrac{26}{15} =$
29. $\dfrac{21}{8} \times \dfrac{16}{9} \times \dfrac{108}{77} =$
30. $\dfrac{7}{3} \times \dfrac{75}{49} \times \dfrac{81}{55} \times \dfrac{49}{9} =$
★31. $\dfrac{9}{5} \times \dfrac{165}{108} \times \dfrac{112}{75} \times \dfrac{90}{77} =$
★32. $\dfrac{18}{11} \times \dfrac{143}{64} \times \dfrac{192}{169} \times \dfrac{65}{36} =$
★33. $\dfrac{572}{14} \times \dfrac{58}{21} \times \dfrac{25}{9} \times \dfrac{43}{16} \times \dfrac{142}{37} \times \dfrac{48}{19} =$

34. A cubic foot of water weighs $62\frac{1}{2}$ pounds. How much do $8\frac{5}{8}$ cubic feet of water weigh?

35. If a plane can fly 700 miles in 1 hour, how far can it fly in $2\frac{3}{4}$ hours?

36. If a city block is $\frac{1}{6}$ mile long, how many miles has a man gone when he has walked $6\frac{3}{4}$ blocks south and $7\frac{1}{2}$ blocks east? If he is trying to walk 5 miles during the day, how many miles does he still lack?

37. If water weighs $62\frac{1}{2}$ pounds per cubic foot, what is the weight of the water in a tank which contains $21\frac{3}{4}$ cubic feet?

38. A recipe reads as follows: 3 cups of flour, 4 teaspoons of baking powder, $\frac{3}{4}$ teaspoon of salt, $2\frac{1}{2}$ tablespoons of sugar, $1\frac{1}{4}$ cups of milk, and $\frac{2}{3}$ cup of shortening. How much of each ingredient will be required to make $5\frac{1}{2}$ times the recipe?

39. If $1\frac{2}{3}$ yards of material is required to make an apron, how many yards are needed to make 7 aprons? If $14\frac{1}{2}$ yards of material were purchased, how much was left over after making 7 aprons?

40. Mortar for 1,000 bricks requires $\frac{5}{8}$ cubic yard of sand and $1\frac{1}{8}$ barrels of lime. If the estimate on a building project calls for 11,350 bricks, how much sand and lime should be ordered?

FRACTIONS _____57

41. The circumference of a circle is approximately $3\frac{1}{7}$ times the diameter. Find the circumference if the diameter is $7\frac{7}{8}$ inches.

★42. The circumference of a circle is approximately $3\frac{1}{7}$ times the diameter. Find the circumference if the diameter is $7\frac{3}{8}$ inches.

43. A feed and seed store estimates that it requires around $2\frac{3}{4}$ bushels of seed wheat per acre. How much seed wheat will be needed for $195\frac{3}{4}$ acres?

44. A cookie recipe requires $1\frac{3}{4}$ cups of sugar. How much sugar will be needed to make $6\frac{1}{2}$ times the recipe?

45. The cooking class is divided into 8 teams. If each team uses a recipe that takes $2\frac{1}{2}$ cups flour, $2\frac{1}{4}$ tea-spoons baking powder, $\frac{1}{2}$ cup shortening, and $1\frac{1}{4}$ cups sugar, how much of each ingredient does the whole class need?

46. If it takes approximately $3\frac{4}{5}$ pounds of feed for one dozen eggs, how much feed will it take for 30 dozen eggs?

6 | Dividing improper fractions and mixed numbers

In dividing two improper fractions we invert the divisor and then multiply the two fractions. For example:

$$\frac{8}{5} \div \frac{56}{15} = \frac{\overset{1}{\cancel{8}}}{\underset{1}{\cancel{5}}} \times \frac{\overset{3}{\cancel{15}}}{\underset{7}{\cancel{56}}} = \frac{3}{7}$$

Here we have $\frac{56}{15}$ as the divisor. Upon inverting $\frac{56}{15}$, we obtain $\frac{15}{56}$. The division process might also be expressed as

$$\frac{\frac{6}{5}}{\frac{126}{25}} = \frac{\overset{1}{\cancel{6}}}{\underset{1}{\cancel{5}}} \times \frac{\overset{5}{\cancel{25}}}{\underset{21}{\cancel{126}}} = \frac{5}{21}$$

In a series of divisions of fractions, such as $\frac{4}{3} \div \frac{9}{8} \div \frac{3}{2}$, we have

$$\frac{4}{3} \times \frac{8}{9} \times \frac{2}{3} = \frac{64}{81}$$

The easiest way to handle the division of mixed numbers is to change them to improper fractions and proceed as just shown in the dividing of improper fractions. For example:

$$17\tfrac{4}{13} \div 6\tfrac{2}{65} = \frac{225}{13} \div \frac{392}{69} = \frac{225}{\underset{1}{\cancel{13}}} \times \frac{\overset{5}{\cancel{65}}}{392} = \frac{1125}{392} = 2\tfrac{341}{392}$$

EXERCISE 7

Divide as indicated:

1. $\dfrac{1}{3} \div \dfrac{1}{4}$ $\dfrac{2}{3} \div \dfrac{1}{3}$ $\dfrac{1}{2} \div \dfrac{3}{2}$ $\dfrac{2}{5} \div \dfrac{3}{5}$ $\dfrac{5}{8} \div \dfrac{5}{6}$

2. $\dfrac{3}{16} \div \dfrac{3}{8}$ $\dfrac{4}{9} \div \dfrac{2}{3}$ $\dfrac{5}{8} \div \dfrac{5}{4}$ $\dfrac{3}{4} \div \dfrac{9}{16}$ $\dfrac{5}{16} \div \dfrac{15}{4}$

3. $\dfrac{9}{4} \div \dfrac{3}{2} =$ 4. $\dfrac{16}{5} \div \dfrac{32}{25} =$ 5. $\dfrac{18}{7} \div \dfrac{54}{49} =$

6. $\dfrac{21}{16} \div \dfrac{105}{64} =$ 7. $\dfrac{10}{3} \div \dfrac{40}{21} =$ 8. $\dfrac{15}{4} \div \dfrac{75}{32} =$

9. $\dfrac{\tfrac{65}{48}}{\tfrac{13}{16}}$ 10. $\dfrac{\tfrac{36}{45}}{\tfrac{9}{5}}$ 11. $\dfrac{\tfrac{18}{25}}{\tfrac{14}{15}}$

12. $\dfrac{\tfrac{54}{25}}{\tfrac{18}{5}}$ 13. $\dfrac{\tfrac{128}{9}}{\tfrac{81}{42}} =$ 14. $\dfrac{\tfrac{76}{11}}{\tfrac{64}{33}} =$

15. $\dfrac{\tfrac{18}{5}}{\tfrac{72}{25}} =$ 16. $\dfrac{\tfrac{256}{9}}{\tfrac{320}{189}} =$ 17. $8\tfrac{2}{5} \div 3\tfrac{3}{5}$

18. $3\tfrac{1}{8} \div 1\tfrac{1}{4}$ 19. $5\tfrac{1}{4} \div 1\tfrac{1}{2}$ 20. $6\tfrac{2}{5} \div 2\tfrac{4}{5}$

21. $5\tfrac{3}{5} \div 2\tfrac{2}{5} =$ 22. $9\tfrac{2}{5} \div 4\tfrac{4}{5} =$

23. $21\tfrac{3}{8} \div 14\tfrac{3}{4} =$ 24. $32\tfrac{3}{16} \div 12\tfrac{5}{8} =$

25. $160\tfrac{3}{4} \div 42\tfrac{3}{16} =$ *26. $256\tfrac{3}{7} \div 48\tfrac{16}{21} =$

27. $6938\tfrac{4}{9} \div 2814\tfrac{6}{7} =$

*28. $\dfrac{3658\tfrac{4}{45}}{2326\tfrac{2}{729}} =$ *29. $\dfrac{2162\tfrac{2}{27}}{538\tfrac{5}{96}}$

30. $\dfrac{8}{3} \div \dfrac{40}{27} \div \dfrac{81}{20} =$ 31. $\dfrac{9}{5} \div \dfrac{63}{140} \div \dfrac{128}{13} =$

32. $\dfrac{35}{6} \div \dfrac{105}{42} \div \dfrac{147}{69} =$ 33. $\dfrac{27}{8} \div \dfrac{243}{128} \div \dfrac{256}{45} \div \dfrac{125}{80} =$

FRACTIONS

34. $\dfrac{48}{9} \div \dfrac{96}{147} \div \dfrac{63}{46} \div \dfrac{92}{87} =$

35. $\dfrac{125}{69} \div \dfrac{165}{46} \div \dfrac{625}{195} \div \dfrac{247}{140} =$

★36. $48\tfrac{3}{4} \div \dfrac{795}{428} \div \dfrac{868}{625} =$

37. $147\tfrac{2}{5} \div \dfrac{171}{60} \div \dfrac{642}{243} =$

★38. $43{,}642\tfrac{3}{17} \div 967\tfrac{3}{5} \div \dfrac{243}{74} \div \dfrac{1428}{729} =$

Perform the indicated operations.

39. $\dfrac{7}{8} \times \dfrac{2}{21} \div \dfrac{16}{45} =$

40. $\dfrac{11}{16} \times \dfrac{8}{33} \div 6\tfrac{3}{8} =$

41. $\dfrac{26}{25} \div \dfrac{2}{5} \times \dfrac{45}{52} =$

42. $\dfrac{169}{15} \times 5\tfrac{1}{13} \div \dfrac{7}{4} =$

43. $\dfrac{225}{12} \div 8\tfrac{3}{4} \times \dfrac{10}{15} =$

44. A piece of metal $12\tfrac{1}{4}$ feet in length is to be divided into 9 equal pieces. What will be the length of each piece if it is necessary to allow one-eighth inch for each cut?

45. Jack Fairchild drives 524 miles in $8\tfrac{3}{4}$ hours. What is the average speed of the car in miles per hour?

46. If a car uses $\tfrac{3}{22}$ of a gallon of gasoline per mile, how many gallons will be required for a trip of 1624 miles?

47. A piece of metal $9\tfrac{3}{4}$ feet in length is to be divided into 9 equal pieces. What will be the length of each piece if it is necessary to allow one-eighth inch for each cut?

48. How many pieces $4\tfrac{5}{16}$ inches in length can be cut from a steel bar $32\tfrac{3}{4}$ inches in length, if one-eighth inch is allowed for each cut?

49. In the blueprint of a house the architect represented 1 foot by three-eighths inch. If a room measures 14 inches by $8\tfrac{3}{8}$ inches on the drawing, what are the dimensions of the room?

50. The diameter of a circle is approximately the circumference divided by $3\tfrac{1}{7}$. Find the diameter if the circumference is $21\tfrac{1}{4}$ feet.

51. How many planks $1\tfrac{5}{16}$ feet long must be laid end to end to make a row 42 feet long?

52. How many pieces of wire each $3\tfrac{3}{4}$ feet long can be cut from a coil of wire 240 feet long?

53. The diameter of a circle is approximately the circumference divided by $3\tfrac{1}{7}$. Find the diameter if the circumference is $16\tfrac{2}{3}$ feet.

54. A steam pipe is to be covered with asbestos. If each piece of the asbestos is $2\frac{3}{4}$ feet long, how many pieces are needed to cover a pipe $46\frac{3}{4}$ feet long?

EXERCISE 8

Review

Add:

1. $\frac{1}{2} + \frac{3}{4} + \frac{5}{6} =$

2. $\frac{1}{3} + \frac{1}{6} + \frac{2}{9} =$

3. $\frac{3}{8} + \frac{2}{3} + \frac{5}{12} =$

4. $\frac{4}{9} + \frac{2}{3} + \frac{5}{27} =$

5. $\frac{1}{2} + \frac{1}{3} + \frac{1}{4} + \frac{1}{6} + \frac{1}{12} =$

6. $\frac{5}{9} + \frac{5}{6} + \frac{7}{12} + \frac{5}{18} =$

★7. $\frac{1}{4} + \frac{2}{3} + \frac{3}{5} + \frac{4}{7} + \frac{5}{6} =$

★8. $\frac{3}{5} + \frac{3}{16} + \frac{5}{8} + \frac{10}{21} + \frac{4}{15} =$

Subtract:

9. $\frac{3}{4} - \frac{1}{2} =$

10. $\frac{5}{6} - \frac{1}{3} =$

11. $\frac{5}{8} - \frac{1}{4} =$

12. $\frac{4}{5} - \frac{3}{10} =$

13. $\frac{7}{8} - \frac{2}{5} =$

Multiply:

14. $\frac{3}{8} \times \frac{4}{9} \times \frac{7}{8} =$

15. $\frac{18}{25} \times \frac{3}{8} \times \frac{3}{4} \times \frac{5}{9} =$

16. $\frac{24}{37} \times \frac{74}{144} \times \frac{16}{27} \times \frac{4}{7} =$

Divide:

17. $\frac{5}{7} \div \frac{5}{8} =$

18. $\frac{5}{18} \div \frac{21}{40} =$

19. $\frac{35}{42} \div \frac{3}{7} =$

FRACTIONS

Add:
20. $6\frac{3}{5} + 4\frac{3}{4} =$
21. $5\frac{1}{2} + 6\frac{1}{4} =$
22. $33\frac{2}{3} + 15\frac{1}{4} + 125\frac{3}{16} =$

Subtract:
23. $86\frac{3}{4} - 35\frac{2}{3} =$
24. $112\frac{4}{15} - 89\frac{11}{30} =$

Multiply:
25. $\dfrac{16}{7} \times \dfrac{49}{4} \times \dfrac{12}{5} =$
26. $\dfrac{18}{5} \times \dfrac{35}{24} \times \dfrac{27}{8} =$
27. $7\frac{1}{2} \times 6\frac{2}{3} \times 3\frac{3}{5} =$
28. $3\frac{3}{4} \times 4\frac{1}{6} \times 2\frac{2}{5} =$
29. $4\frac{3}{8} \times 5\frac{3}{5} \times 3\frac{1}{7} =$
30. $8\frac{1}{6} \times 5\frac{1}{7} \times 4\frac{2}{3} =$
★**31.** $17\frac{1}{2} \times 7\frac{2}{3} \times 19\frac{1}{4} =$
★**32.** $142\frac{1}{4} \times 63\frac{3}{8} \times 42\frac{3}{4} =$

Divide:
33. $\dfrac{125}{13} \div \dfrac{440}{91} =$
34. $\dfrac{145}{14} \div \dfrac{365}{112} =$
35. $96\frac{1}{4} \div 27\frac{1}{2} =$
36. $125\frac{5}{8} \div 33\frac{3}{4} =$
37. $85\frac{2}{3} \div 15\frac{1}{4} =$

38. Raymond planted $\frac{3}{4}$ of a 16-acre field with corn. How many acres of corn did he plant?

39. Alice bought five sheets of postage stamps (100 to a sheet) and used three-fourths of them the first week. How many did she have left?

40. Jack reads 300 words a minute and Jim reads three-fifths as many. How many words a minute does Jim read?

41. If potatoes are four-fifths water, how many pounds of water are there in 10 bushels? (Assume that a bushel of potatoes weighs 60 pounds.)

42. 600 feet is an increase of $\frac{4}{5}$ over ? feet?

43. On a certain line of goods the net price is one-fourth less than the list price. If the net price of an article is $3.60, what is its list price?

44. A cost-price formula requires the cost of manufacture to be two-thirds less than the selling price. Find the selling price if the cost is $5.

45. If, on the plan of a house that the Maxwells are studying, $\frac{1}{8}$ inch represents 1 foot, what are the dimensions of a room that measures $1\frac{3}{4}$ inches by $1\frac{7}{8}$ inches on the plan?

46. A 6-pound roast lost one-fourth of its weight in cooking. How much did it weigh after cooking?

47. Maurice worked $3\frac{1}{4}$ hours last night on his story for the school paper. Today he finished it in $1\frac{1}{2}$ hours. How long did it take Maurice to write the story?

48. A share in the catch of the Betz Fishing Schooner is one-fortieth of the value of the catch (after certain fixed charges are met). Bill's allotment is $4\frac{3}{8}$ shares. To what part of the value of the catch is Bill entitled?

49. Can you divide 7 dollar bills among 8 men so that each gets 1 dollar?

50. A lifeboat carries 6 gallons of water. Five persons are in the lifeboat. What part of a gallon of water should each be allowed?

51. Best Yet Hamburger Stand buys ground beef in 80-pound lots. If $\frac{2}{9}$ of a pound of ground beef goes into each of their hamburgers, how many hamburgers can they sell from 80 pounds of ground beef?

52. A box contains 8 oranges. Can you divide these among 8 boys, in such a way as to leave 1 orange in the box? (None of the oranges should be cut.)

53. If vegetables, when dried, lose $\frac{9}{14}$ of their fresh weight, how much will $106\frac{2}{3}$ pounds of fresh vegetables weigh when dried?

54. By what must 20 be divided to yield a quotient of $3\frac{3}{5}$?

55. A 1-2-4 concrete mixture means 1 part cement, 2 parts sand, and 4 parts gravel. What fractional part of the total mixture is sand?

56. James paid $187 for a ring, which included the price of the ring and a tax of one-tenth of this price. What was the price of the ring before taxes?

57. Bemis buys a piece of luggage for $51.70. This price includes the cost of the luggage and a tax of one-tenth of this cost. What was the price of the luggage before the tax was added?

58. Jake wishes to buy a record player priced at $80. He pays $\frac{1}{5}$ in cash and the rest in 8 equal monthly installments. How much must he pay each month?

Chapter 4

Decimals

1 | Extension of place value—decimals

As we observed in Chapter Two, the notation for the natural numbers 0, 1, 2, 3, 4, 5, 6, 7, 8, 9, is also called the *decimal system* (from the Latin *decem*, meaning ten) because ten symbols or digits are used to represent all numbers.

We have already discussed the fact that the place value given to a digit depends on the place the digit occupies relative to the units place. We have illustrated this fact with examples in which the digits were to the left of the units place and in the units place. Thus 5943 means $5(1000) + 9(100) + 4(10) + 3(1)$ or $5(10)^3 + 9(10)^2 + 4(10)^1 + 3(10)^0$.

Let us place a period or dot (.), called the *decimal point*, at the right of the units place and explain place value as it operates at the right of the decimal point. A digit written immediately to the right of the decimal point has a place value which is one-tenth of the place value the digit would have if it stood in the units place.

Let us refer to Figure 4 in which the digit 6, appearing immediately to the right of the decimal point, represents six-*tenths*, or $\frac{6}{10}$. A digit in this position is described as being in the *tenths place*. A digit immediately to the right of the tenths place has a place value equal to one-tenth of the place value the digit would have if it were in the tenths place or one-hundredth of the place value it would have if it were in the units place. In Figure 4, the digit 5 appearing in the second place to the right of the decimal point stands for five *one-hundredths*, or $\frac{5}{100}$. A digit in this position is described as being in the *hundredths place*.

In accordance with the principle of place value, the place value of a digit (both to the left and to the right of the decimal point) is 10 times the place value which that digit would have if it stood in the next place to the right. Digits on the right of the decimal point in a number stand for the number of tenths, hundredths, thousandths,

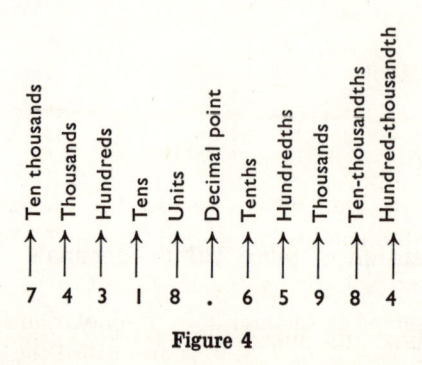

```
            Ten thousands
              Thousands
                Hundreds
                  Tens
                    Units
                      Decimal point
                        Tenths
                          Hundredths
                            Thousands
                              Ten-thousandths
                                Hundred-thousandth
              ↑  ↑  ↑  ↑  ↑  ↑  ↑  ↑  ↑  ↑
              7  4  3  1  8  .  6  5  9  8  4
```

Figure 4

and so forth, that appear in the number. We read the number 74318.65984 in Figure 4 as "seventy-four thousand, three hundred eighteen *and* sixty-five thousand, nine hundred eighty-four hundred-thousandths." In reading a decimal number, the word "and" is used only for the decimal point. Further examples are:

6.7 is read "six and seven-tenths"
18.03 is read "eighteen and three-hundredths"
4634.232 is read "four thousand six hundred thirty-four and two hundred thirty-two-thousandths."

Decimal places refer to the places held by digits in a decimal to the right of the decimal point. Thus 63.75 has two decimal places and 124.687 has three decimal places.

A number that has digits other than zero to the right of the decimal point is called a decimal. Thus 643.29 and 0.831 are decimals. Let us observe that the part of a decimal to the left of the decimal point represents an integer or whole number. Let us examine the part of a decimal to the right of the decimal point. It is observed to be a proper fraction whose denominator is some power of 10. This

DECIMALS 67

fraction is also called a decimal fraction. Thus $0.7 = \frac{7}{10}$, $0.07 = \frac{7}{100}$, and $0.073 = \frac{73}{1000}$ are all proper fractions whose denominator is some power of 10. $\frac{7}{10}$ means the same thing as $\frac{7}{10^1}$, $\frac{7}{100}$ means the same thing as $\frac{7}{10^2}$, and $\frac{73}{1000}$ has the same value as $\frac{73}{10^3}$. Now let us reconsider the decimal referred to in Figure 4, namely, 74318.62984. It may be rewritten as

$$7(10{,}000) + 4(1000) + 3(100) + 1(10) + 8(1)$$
$$+ 6\left(\frac{1}{10}\right) + 5\left(\frac{1}{100}\right) + 9\left(\frac{1}{1000}\right) + 8\left(\frac{1}{10{,}000}\right) + 4\left(\frac{1}{100{,}000}\right)$$

This example shows again the positional nature of the decimal system which we treated to some extent in Chapter Two, "Whole Numbers."

2 | Adding and subtracting decimal fractions

We have already observed in connection with the addition and subtraction of whole numbers that only items of the same kind can be added and subtracted. It does not seem unreasonable therefore for us to say that tenths must be combined with tenths, hundredths with hundredths, thousandths with thousandths, and so forth, in adding and subtracting decimals.

It is important in both adding and subtracting decimals to keep the decimal points in a line, as well as to have all units (and likewise all tens, hundreds, tenths, hundredths, thousandths, and so forth) in a separate column.

For example add:

$$\begin{array}{r} 284.76 \\ 78.928 \\ 64.09 \\ 12.715 \\ \hline 440.493 \end{array}$$

Here we have units, tens, and hundreds in individual columns as well as tenths, hundredths, and thousandths.

Starting the addition with the extreme right-hand column and working to the left by columns, we obtain the following results, *in terms of place value.*

$$\frac{8}{1000} + \frac{5}{1000} = \frac{13}{1000} = \frac{10}{1000} + \frac{3}{1000} = \frac{1}{100} + \boxed{\frac{3}{1000}}$$

$$\frac{1}{100} + \frac{6}{100} + \frac{2}{100} + \frac{9}{100} + \frac{1}{100} = \frac{19}{100} = \frac{10}{100} + \frac{9}{100} = \frac{1}{10} + \boxed{\frac{9}{100}}$$

$$\frac{1}{10} + \frac{7}{10} + \frac{9}{10} + \frac{0}{10} + \frac{7}{10} = \frac{24}{10} = \frac{20}{10} + \frac{4}{10} = 2 + \boxed{\frac{4}{10}}$$

$$2 + 4 + 8 + 4 + 2 = 20 = 2(10) + \boxed{0(1)}$$

$$2(10) + 8(10) + 7(10) + 6(10) + 1(10) = 24(10)$$
$$= 20(10) + 4(10) = 2(100) + \boxed{4(10)}$$

$$2(100) + 2(100) = \boxed{4(100)}$$

The entire answer can be diagrammed as:

$4(100) + 4(10) + 0(1) + 4(0.1) + 9(0.01) + 3(0.001)$, or
$4(10^2) + 4(10^1) + 0(10^0) + 4(10^{-1}) + 9(10^{-2}) + 3(10^{-3})$.

An example in subtraction follows:

$$\begin{array}{r} 6547.34 \\ 168.274 \\ \hline 6379.066 \end{array}$$

Here the borrowing of the powers of 10 is similar to the carrying procedure just given for addition.

Common fractions may be converted to decimal fractions by dividing the denominator into the numerator.

Thus the fraction $\frac{3}{8}$ becomes the decimal fraction 0.375, when we divide the 8 into the 3, and the fraction $\frac{5}{4}$ becomes the decimal fraction 1.25 when we divide the 4 into the 5. Other examples of converting common fractions into decimal fractions are: $\frac{3}{5} = 0.6$, $\frac{5}{16} = 0.3125$, $\frac{7}{4} = 1.75$, and $\frac{9}{5} = 1.8$.

Decimal fractions likewise may be changed into common fractions. For example, 0.125 means $\frac{125}{1000}$, which reduces to $\frac{1}{8}$; $0.04 = \frac{4}{100} = \frac{1}{25}$; $2.6 = 2\frac{3}{5} = \frac{13}{5}$; $7.125 = 7\frac{125}{1000} = 7\frac{1}{8} = \frac{57}{8}$; and $8.5 = 8\frac{5}{10} = 8\frac{1}{2} = \frac{17}{2}$.

DECIMALS _____69

EXERCISE 1

Add:

1. 0.5 0.2 0.1	**2.** 0.5 0.4 0.9	**3.** 0.04 0.01 0.03
4. 0.16 0.07 0.54 0.13	**5.** 0.38 0.46 0.73 0.85	**6.** 0.97 0.12 0.04
7. 2.8 3.1 5.6	**8.** 7.3 6.4 5.5 7.6	**9.** 5.08 7.09 3.07
10. 7.63 4.98 7.64	**11.** 4.207 5.764 7.658	**12.** 35.47 52.69 47.83
13. 438.09 192.68 348.79 583.38	**14.** 18.63 7.58 486.79 65.34 137.65 82.44	
15. 43.67 17.92 8.67 85.36 0.84	**16.** 496.38 715.27 809.63 690.44 187.73	

17. $0.5 + 0.3$

18. $0.07 + 0.16$

19. $5.7 + 8.6$
20. $0.362 + 0.47$
21. $0.63 + 0.492$
22. $0.68 + 0.1459$
23. $21 + 0.57$
24. $0.023 + 17$
25. $0.87 + 0.025$
26. $0.1783 + 0.39$
27. $0.07 + 8$
28. $0.736 + 0.37$
29. $0.76 + 8 + 5.36 + 0.076$

Subtract:

30. 0.9
     0.4

31. 0.29
     0.16

32. 0.85
     0.43

33. 0.08
     0.03

34. 0.97
     0.63

35. 0.64
     0.37

36. 5.7
     3.4

37. 0.875
     0.392

38. 4.83
     1.86

39. 17.8
     4.6

40. 0.7863
     0.3854

41. 82.695
     38.486

42. 0.49643
     0.15628

43. 0.97643
     0.85384

44. $0.18 - 0.07$
45. $0.375 - 0.2$
46. $0.9 - 0.675$
47. $3.6 - 0.72$

Add the following:

★48.   476.83
         39.367
       1892.487
       5368.684
      39642.3
          0.0067
       9463.734

★49. 9643.888
         37.36
          8.237
        359.583
       9846.87
     14379.8642
       869.003

DECIMALS

*50.	9763.47	*51.	70000.008
	759.886		506.2793
	5493.6576		8923.6734
	92.307		66666.5555
	864.7538		5842.3178
	64799.8989		50709.0206
	3007.0096		9863.44487
	7400.5203		

Subtract the following and check by adding the remainder to the subtrahend.

52. 9763.857
    306.005

53. 35983.4063
    8864.759

54. 8777.8643
    609.6096

55. 9638.753
    459.876

Write the following numbers in words.

56. 649.37
57. 486.93
58. 8765.764
59. 9763.789
60. 15,579.5896
61. 592,648.7963
62. 753,462.9078
63. 709,547.6302
*64. 9,793,652.68347
*65. 59,756,987.67834

Express the following in numbers.

66. Nine thousand, five hundred thirty-six and fifty-nine hundredths.

67. Three thousand, eight hundred forty-six and seventy-three hundredths.

68. Eighty-nine thousand, eight hundred sixty-four and five hundred ninety-two thousandths.

69. Forty-five thousand, six hundred forty-seven and eight hundred twenty-nine thousandths.

*70. Nine hundred fifty-four thousand, six hundred thirty-three and nine thousand five hundred sixty-eight ten thousandths.

*71. Three hundred ninety-four thousand, four hundred sixty-nine and six thousand nine hundred seven-ten thousandths.

*72. Six million, four hundred eighty-two thousand, six hundred eighty-three and fifty-eight thousand eight hundred fifty-five hundred thousandths.

*73. Five hundred forty-five million, six hundred seventy-five thousand, five hundred thirty-seven and sixty-seven thousand four hundred sixty-seven hundred thousandths.

Add the following:

*74. 65.37 + 842.679 + 3692.577 + 502.008

*75. 642.88 + 8937.06 + 7005.4967 + 63.54

*76. 73,695.8467 + 42.39 + 548.6608 + 542,008.86592 + 0.5842 + 7963.259

*77. 596,493.037 + 15,709.6875 + 42.38657 + 59.36945 + 538.576 + 8547.5907

78. Mrs. Branch has a credit card at a department store. During a certain month she charged items for the following amounts: $5.64, $25.93, $4.63, $154.69, $92.76, $41.52, $15.96, $63.49, and $142.63. At the beginning of the month she already owed $19.57. During the month she makes three payments of one hundred fifty dollars each and one payment of one hundred dollars. How did her account stand at the end of the month?

79. At the beginning of the accounting period for a certain month, the Blanchard family checking account has a balance of $225.62. During this accounting period the following deposits are made: $510.43, $46.97, $453.64, and $329.58. Checks for the following amounts are written: $11.62, $149.63, $19.47, $78.63, $34.55, $15.48, $38.77, $85.29, $9.65, $45.77, $128.65, $49.57, $42.33, $128.44, $5.07, $39.66, $439.88, $68.03, and $43.29. Find the balance in the account at the close of the accounting period.

80. (a) How much should a dealer sell a television set for if it cost him $98.75 and he wishes to make a profit of $32.50? (b) Jack paid $48.55 for a bicycle. What was the original price if the store manager gave him a reduction of $8.50?

81. The utilities bills in the Prescott Apartments run as follows for 12 months: $596.48, $643.28, $629.35, $538.84, $694.39, $704.69, $657.49, $624.76, $618.59, $605.85, $587.64, and $598.38. Find the total for the year.

*82. Mel goes 10 miles per day, Jack 12, Ralph 16, Sam 24, and Ronald 30. On June 1, they all start simultaneously going in the

DECIMALS

same direction around a circular island 56 miles around. When will they all meet?

Add:

83. $54 + 0.076 + 2.34 + 0.009 + 16.309 =$
84. $0.96 + 395 + 0.493 + 57 + 0.07 =$

Subtract:

85. $769 - 0.593 =$ **86.** $1692 - 0.047 =$
87. $31 - 0.973 =$ **88.** $793 - 0.704 =$

Change the following to decimal fractions.

89. $\frac{4}{5}, \frac{9}{10}, \frac{7}{8}, \frac{9}{16}$, and $\frac{3}{15}$ **90.** $\frac{1}{2}, \frac{3}{4}, \frac{5}{8}, \frac{3}{16}$, and $\frac{5}{32}$

91. $\frac{6}{4}, \frac{9}{8}, \frac{19}{4}, \frac{26}{5}$, and $\frac{25}{8}$ **92.** $\frac{5}{2}, \frac{7}{4}, \frac{7}{5}, \frac{15}{2}$, and $\frac{17}{8}$

Change the following decimal fractions to common fractions.

93. 0.36, 0.425, 0.625, 0.96, and 0.0625
94. 0.3125, 0.555, 0.6, and 0.0025
95. 1.2, 45.35, 16.625, 21.375, and 140.25
96. 2.5, 7.8, 9.125, 3.0625, and 25.875

3 | Multiplying decimal fractions

Multiplication of decimal fractions is carried on in a manner similar to that used with whole numbers and described in Chapter Two. When we multiply 47.63 by 32.57 we line the numbers up with hundredths under hundredths, tenths under tenths, units under units, tens under tens, and hundreds under hundreds. Again it is important to observe place value.

Thus:

$$
\begin{array}{r}
47.63 \\
\times\ 32.57 \\
\hline
\end{array}
\quad \text{is equal to} \quad
\begin{array}{r}
47.63 \\
\times\ 0.07 \\
\hline
3.3341
\end{array}
+
\begin{array}{r}
47.63 \\
\times\ 0.50 \\
\hline
23.8150
\end{array}
$$

$$
+
\begin{array}{r}
47.63 \\
\times\ 2.00 \\
\hline
95.2600
\end{array}
+
\begin{array}{r}
47.63 \\
\times\ 30.00 \\
\hline
1428.9000
\end{array}
$$

Hence we have:

$$\begin{array}{r} 47.63 \\ \times\ 32.57 \\ \hline 3.3341 \\ 23.8150 \\ 95.2600 \\ 1428.9000 \\ \hline 1551.3091 \end{array}$$

In each of the separate products, 47.63×0.07, 47.63×0.50, 47.63×2.00, 47.63×30.00, we are multiplying hundredths by hundredths, which gives ten-thousandths. Each time, consequently, we point off four places. For example:

$$0.23 \times 0.15 = \frac{23}{100} \times \frac{15}{100} = \frac{23 \times 15}{100 \times 100} = \frac{345}{10000} = 0.0345$$

A simple rule for multiplying decimal fractions is to take the sum of the number of decimal places in the multiplier, and the number of decimal places in the multiplicand and point off this number of decimal places in the product.

Thus, in the example 0.23×0.15 we take the sum of two decimal places in the multiplier and two decimal places in the product, and obtain a total of four decimal places. We point off four decimal places in the product, and we have 0.0345.

EXERCISE 2

Multiply:

1. 0.4 9	**2.** 75 0.23	**3.** 0.471 93	**4.** 0.7934 6
5. 86 0.43	**6.** 45 0.07	**7.** 0.08 3	**8.** 19 0.005
9. 0.007 5	**10.** 3.64 19	**11.** 0.8 0.3	**12.** 0.5 0.6

DECIMALS

13. 0.72 0.9	14. 0.37 0.5	15. 0.67 0.83	16. 0.07 0.03
17. 15.3 0.047	18. 37.65 0.863	19. 0.246 0.09	20. 3.1416 0.85
21. 0.063 0.047	22. 0.013 × 0.008		23. 38.65 29.6
24. 84.36 39.5	25. 929.3 4.67		26. 857.65 0.926
27. 693.572 8.35	28. 874.685 94.2		29. 4932.78 68.3
30. 9763.45 0.768	*31. 784.56 685.7		*32. 53.76 79.893

*33. 68.79
8765.4

34. Find the cost of 24 cans of fruit @ $0.39 each.

35. Find the cost of 3¾ pounds of meat @ $0.98 per pound.

36. Find the cost of 4 tires @ $22.95 each.

37. A refrigerator costs $385 cash or $38.50 down and 12 payments of $30.87 each. How much is saved by paying cash?

38. An airplane has a ground speed of 625 knots. What is its ground speed in statute mph if 1 knot = 1.15 statute miles per hour?

39. Find the distance represented by 8.3 inches if the scale is 1 inch = 50 miles.

40. A vessel heads N 18° W for 9 hours at 10.8 knots. Find the distance traveled in nautical miles. (1 knot = 1 nautical mph)

41. The Dragon Baseball team in the Pony League bought 12 balls at $1.95 each and 9 bats at $2.15 each. Find the total spent.

42. The rates of a local laundry are $1.55 for the first 12 pieces and 0.11 for each additional piece. If 25 pieces are laundered, find the total charge.

43. The telephone rates between 2 zones is 0.85 for the first 3 minutes and 0.17 for each additional minute. What is the charge on a call that lasts 7 minutes?

44. The hold of a freighter contains 40,000 bushels of potatoes. If one bushel equals 1.25 cubic feet, how many cubic feet of potatoes are on the freighter?

45. The board of directors of an airline needs to provide money for 46 new jet planes. It finds that the cost of these jets is $26,543,960 each. How much money will need to be secured to finance this purchase?

46. The manager of the Fashion Clothing Store finds that if he buys 160 suits the price will be $52.80 each, whereas if he buys 260 suits the price will be $46.50 each. How much additional money is needed by the manager if he purchases 260 suits rather than 160 suits?

47. If runways cost $2846.50 a foot, how much will it cost to build 120.4 miles of new runways for jet planes? (There are 5280 feet in a mile.)

48. A rectangular field is 649.34 feet long and 406.93 feet wide. Find the total distance around the field.

49. A food purchase at Quik-Way Grocery includes the following items: 6 pounds of rump roast at $1.12 per pound; 3 pounds of center-cut ham at $1.89 per pound; and 9 pounds of hamburger at $1.03 per pound. Canned goods are included as follows: ten cans at 2 for 47 cents; 8 cans at 32 cents per can; 6 cans at 2 for 89 cents; and 4 cans at 43 cents per can. The checker was given two twenty-dollar bills. How much change did the customer receive?

50. The Collins Market buys 186 pounds of dressed hens at 46 cents per pound and sells them at 58 cents per pound. Find the profit on the entire transaction.

51. Mr. Michael buys 178 bushels of sweet potatoes at $1.86 per bushel and sells them at $3.65 per bushel. Find his profit on the whole transaction.

52. Find the total cost of 20 pounds of sugar at $0.16 per pound, 4 dozen oranges at $0.52 per dozen, and 3 pounds of coffee at $1.12 per pound.

53. At $365 per acre, what is the total cost of six tracts of land that contain respectively 8.5 acres, 16.9 acres, 42.3 acres, 5.7 acres, 86.2 acres, and 123.9 acres?

DECIMALS _____77

54. Mr. Robinson receives $1.367 per hour for 38.7 hours per week. Mr. James receives $1.264 per hour for 42.6 hours per week. Which man earns more per week? How much more does he earn than the other worker?

4 | Dividing decimal fractions

Division of decimal fractions is handled in a way similar to that used in Chapter Two with whole numbers. When we divide 6.25 by 2.5, we have $2.5\overline{)6.25}$. We recall that the 2.5 is called the divisor and the 6.25 is called the dividend. In dividing decimal fractions we proceed to transform the divisor into a whole number by multiplying it by 10, or 100, or 1000, or whatever power of 10 it takes to change the divisor into a whole number. By a power of 10 we recall that 10^2 means 10×10, or 100, 10^3 means $10 \times 10 \times 10$, or 1000, 10^4 means $10 \times 10 \times 10 \times 10$, or 10,000, etc. In the present illustration,

$$2.5\underset{\smile}{.}\overline{\big)6\underset{\smile}{.}2.5}$$
$$\underline{50}$$
$$125$$
$$125$$

we observe that if we multiply the divisor, 2.5, by 10, we obtain the whole number 25. We also multiply the dividend or 6.25 by 10, obtaining 62.5. Now immediately place the decimal point in the answer (quotient) directly above its new position in the dividend. Division here is carried out in the same manner as division of integers. Upon dividing 25 into 62.5, we obtain 2.5 as the quotient. Let us notice here that when we multiply both the numerator and denominator of a fraction by the same number we do not change the value of the fraction. Thus $\frac{2}{5}$ becomes $\frac{4}{10}$ when both numerator and denominator are multiplied by 2. Two-fifths becomes $\frac{6}{15}$ when both numerator and denominator are multiplied by 3. And $\frac{2}{5}$ becomes $\frac{20}{50}$ when both numerator and denominator are multiplied by 10. Each of the fractions $\frac{4}{10}$, $\frac{6}{15}$, and $\frac{20}{50}$ can be reduced to $\frac{2}{5}$ by dividing both numerator and denominator by the appropriate number. Thus in $\frac{4}{10}$, when both numerator and denominator are divided by 2, we have $\frac{2}{5}$; in $\frac{6}{15}$, when both numerator and denominator are divided by 3, we have $\frac{2}{5}$; and in $\frac{20}{50}$, when both numerator and denominator are divided by 10, we have $\frac{2}{5}$.

Again referring back to the illustration $2.5\overline{)6.25}$, or $\frac{6.25}{2.5}$, or $\frac{62.5}{25}$, we see that 25 divides into 62 two times but not three times, and the answer apparently falls between 2 and 3. As we have just recently shown, the answer is 2.5.

Discussion of approximate numbers will appear in a later chapter. However, in the present chapter we shall set up a partial procedure for the "rounding" of numbers.* For example, "round" 87.688, 87.684, 87.685, and 87.675 to two decimal places each. If the digit to be "rounded" is followed by a digit greater than 5, we add one to the digit to be "rounded." Thus, 87.688 becomes 87.69. If the digit to be "rounded" is followed by a digit less than 5, we leave the digit to be "rounded" unchanged. Thus, 87.684 becomes 87.68. If the digit to be "rounded" is even and is followed by a 5, we leave the digit to be "rounded" unchanged. Then 87.685 becomes 87.68. If the digit to be "rounded" is odd and is followed by a 5, we add one to the digit to be "rounded." Thus, 87.675 becomes 87.68. Let us follow these practices as we work with the division of decimal fractions in this chapter. For example:

$$\begin{array}{r}19.835\\82.47\overline{)1635.79.462}\\8247\\\hline 81109\\74223\\\hline 68864\\65976\\\hline 28886\\24741\\\hline 41452\\41235\\\hline\end{array}$$

---

* In a specific job situation instructions for rounding may differ somewhat from those listed in the text.

DECIMALS

Dividing 1635.79462 by 82.47, carrying the result to three decimal places, and then "rounding" to two decimal places, we obtain 19.84.

EXERCISE 3

Round to the nearest tenth:
1. 0.13
2. 0.34
3. 1.74
4. 6.472
5. 7.304
6. 0.35
7. 0.67
8. 0.874
9. 5.27
10. 4.86

Round to the nearest hundredth:
11. 0.433
12. 0.842
13. 7.576
14. 9.7036
15. 16.0713
16. 0.718
17. 0.407
18. 6.947
19. 8.6798
20. 17.3462

Round to the nearest thousandth:
21. 0.6453
22. 0.8394
23. 7.0497
24. 3.0096
25. 37.78304
26. 0.3257
27. 0.5885
28. 5.0927
29. 3.6839
30. 17.85736

Round each of the following to two decimal places.
31. 87.493
32. 64.387
33. 136.785
34. 249.895
35. 1642.678
36. 2753.683
37. 3499.735
38. 24.365
39. 36.758
40. 147.682

Round each of the following to three decimal places.
41. 607.7853
42. 816.7656
43. 326.8457
44. 1977.2354
45. 2364.9555
46. 7562.9545
47. 927.2625
48. 87.3575
49. 64.2968
50. 267.3874
51. 3618.5942
52. 4723.6879
53. 5783.6755
54. 7932.7885

In Problems 55 through 66, divide, carry the result to three decimal places, and round to two decimal places unless the division comes out even before the third decimal place is reached.

55. $3.56\overline{)7.93765}$

56. $46.7\overline{)95.6984}$

57. $896\overline{)854.792}$

58. $9.63\overline{)7489.54379}$

59. $753.4\overline{)17.79364}$

60. $0.035\overline{)146.7937}$

61. $0.069\overline{)0.008463}$

62. $6.9\overline{)0.7983}$

63. $894.3\overline{)61.79367}$

64. $0.00692\overline{)0.00567395}$

65. $0.6937\overline{)23.6954896}$

66. $7.5\overline{)0.00069532}$

67. The Brown Beaver Scout Troop collected 1786 pounds of scrap paper, which they sold for 92 cents per 100 pounds. What did they receive for the paper?

★68. The unit of electric energy is the watt. A 100-watt light bulb consumes 100 watt-hours of electricity in 1 hour. A kilowatt-hour (kwh) is equal to 1000 watt-hours. The Booster Club rents a church recreation hall for a "fun night." The club agrees to pay $16 plus the cost of the lights. In the recreation hall there are one hundred twenty-four 150-watt light bulbs and sixty-four 200-watt bulbs. What is the rental if the club will require the use of the hall for 9 hours and 15 minutes, and electric energy costs 4.5 cents per kwh?

69. A farmer sells 268 bushels of wheat at $1.86 per bushel and takes his pay in flour at 7.25 cents per pound. How many 100-pound sacks of flour does he receive?

70. The K and S Feed Store buys a carload of wheat for $48.00 a ton and sells it for $2.22 a bushel. If the wheat weighs 8640 pounds, what profit does the feed store realize? (Assume that 1 bushel wheat = 60 pounds.)

71. A ship sails 460 nautical miles. (A nautical mile is 1.15 land or statute miles.) How many statute miles does the ship sail?

72. The manager of Fagan Brothers exchanges 456 yards of silk purchased at $2.14 per yard for percale at 14.4 cents per yard. How many yards of percale does he receive?

73. A metal alloy used to make bearings for certain machinery is 0.73 copper, 0.13 tin, and 0.14 zinc. How many pounds of each metal are in 186 pounds of alloy?

74. How much must be paid for 174 feet of steel bar weighing 1.8 pounds per foot and costing $9.84 per hundred pounds?

75. A knot is a nautical mile per hour. A ship has a speed of 42 miles per hour. What is its speed in knots? Refer to Problem 71.

76. If 1 quart of paint will cover 238 square feet, how many quarts will be needed to cover 14 walls, each 21 feet long and 8 feet high? Give the answer to the nearest quart.

77. How can 10 figs be placed in 3 cups so that every cup will contain an odd number of figs?

78. The athletic association ticket sales this year at the Tarbox School were distributed as follows: 7A, $32.75; 7B, $48.50; 8A, $57.25; 8B, $95.00; 9A, $42.75; and 9B, $34.75. What was the total amount of sales?

79. The net worth of a business is equal to its assets minus its liabilities. Find the net worth of the Black and Sons Hardware whose assets are cash, $1285.75; merchandise, $14,654; accounts receivable, $742.65; other assets, $1950, and whose liabilities are notes payable, $600; accounts payable, $124.62; and other liabilities, $428.84.

Chapter 5

Weights and Measures

1 | National system of weights and measures

The history of our units of measure is quite interesting. Some of these units were of natural origin. The length of the foot was, to begin with, the actual length of the human foot. It varied from $9\frac{3}{4}$ inches to 19 inches at different times according to English history. It is said that Henry I, who lived in the twelfth century, proclaimed that the lawful yard was the distance from the end of his nose to the end of his thumb when his arm was outstretched horizontally. The mile was originated by the Romans. They used it as the length of 1000 double paces where the pace was approximately $2\frac{1}{2}$ feet long. Henry VIII in the sixteenth century set up the weight of the pound as the weight of 7000 grains of wheat taken from the middle of the ear and well dried. The early units of measure were indefinite and variable and led to considerable confusion. They were, of course, a handicap in setting up trade agreements between various countries. As civilization moved forward it was necessary to adopt standard units of measure.

The standard unit of measure needs to be rigidly defined in terms of some unchanging element of nature in order that the unit may be rebuilt in various localities. In 1826, it was decided in England that the imperial yard would be $\frac{360,000}{391,393}$ of the length of a seconds pendulum at sea level in the latitude of London. After a great deal of work, the length was determined, and the standard yard was placed in the House of Parliament. Of course errors were made in the measurements. Also the standard was destroyed when the House of Parliament burned in 1834. The Astronomical Society had made a copy of the standard and it became the basis for the English imperial yard, which is now defined as the distance between two scratches on a special metal rod at a given temperature.

The French government undertook to develop an ideal unit of length early in the nineteenth century. They defined the meter to be $\frac{1}{10,000,000}$ of the distance from one pole of the earth to the equator. This was determined after a very extensive survey and the standard meter was set up. However, errors were made in the calculations, and actually the meter has no special relation to the length of the earth's quadrant. Instead, the standard meter is a completely arbitrary and artificial length defined as the distance at the temperature of melting ice between the centers of two lines etched on the platinum iridium bar deposited at the International Bureau of Weights and Measures, Paris, France.

After a standard unit of length has been derived, the standard units of weight and volume can be expressed in terms of this unit of length. The English troy pound, for instance, equals the weight of 22.778 cubic inches of water at 4°C. The United States gallon is the equivalent of 231 cubic inches. The gram, which is the basic unit of weight in the metric system, equals the weight of one cubic centimeter of distilled water at 4°C.

The United States Bureau of Standards in Washington has a copy of the standard meter, and Congress has defined the fundamental units of our national system of weights and measures in terms of this standard meter. For example, Congress has said that the national yard shall be equal to $\frac{3600}{3937}$ of a meter, or 39.37 inches.

Listed here are the common measures used in the United States for length, area, volume, and capacity.

LENGTH

12 inches = 1 foot
3 feet = 1 yard
$16\frac{1}{2}$ feet $\Big\} = 1$ rod
$5\frac{1}{2}$ yards

5280 feet
1760 yards $\Big\} = 1$ mile
320 rods

WEIGHTS AND MEASURES ———————————85

AREA

144 square inches = 1 square foot
9 square feet = 1 square yard
$30\frac{1}{4}$ square yards = 1 square rod

160 square rods = 1 acre
640 acres = 1 square mile

VOLUME

1728 cubic inches = 1 cubic foot
27 cubic feet = 1 cubic yard

DRY MEASURE

2 pints = 1 quart
8 quarts = 1 peck

4 pecks = 1 bushel
1 bushel = 2150.42 cubic inches

LIQUID MEASURE

2 gills = 1 cup
2 cups = 1 pint
2 pints = 1 quart
4 quarts = 1 gallon

1 gallon = 231 cubic inches
$31\frac{1}{2}$ gallons = 1 barrel
2 barrels = 1 hogshead

AVOIRDUPOIS WEIGHT

$\left.\begin{array}{l}7000 \text{ grains} \\ 16 \text{ ounces}\end{array}\right\} = 1 \text{ pound}$

100 pounds = 1 hundredweight

2000 pounds = 1 ton
2240 pounds = 1 long ton

APOTHECARIES' WEIGHT

20 grains = 1 scruple
3 scruples = 1 dram
8 drams = 1 ounce

$\left.\begin{array}{l}12 \text{ ounces} \\ 5760 \text{ grains}\end{array}\right\} = 1 \text{ pound}$

APOTHECARIES' LIQUID

60 minims = 1 fluid dram
8 fluid drams = 1 fluid ounce

16 fluid ounces = 1 pint
8 pints = 1 gallon

TIME

60 seconds = 1 minute
60 minutes = 1 hour
24 hours = 1 day
7 days = 1 week
52 weeks = 1 year

365 days = 1 common year
366 days = 1 leap year
360 days = 1 commercial year
10 years = 1 decade
100 years = 1 century

SURVEYORS' LENGTH

7.92 inches = 1 link
25 links = 1 rod
4 rods = 1 chain
80 chains = 1 mile

SURVEYORS' AREA

625 square links = 1 square rod
16 square rods = 1 square chain
10 square chains = 1 acre
640 acres = 1 section
36 sections = 1 township

PAPER

24 sheets = 1 quire
20 quires = 1 ream
2 reams = 1 bundle
5 bundles = 1 bale
500 sheets = 1 ream

MISCELLANEOUS

1 acre = 40 yards × 120 yards, approximately
1 carat = 200 milligrams
1 cord = 128 cubic feet
1 hand = 4 inches
1 furlong = 40 rods
1 cubic foot = about $7\frac{1}{2}$ gallons
1 knot = 1.15 miles per hour

Avoirdupois weight is commonly used for weighing objects. The long ton is used at times where coal and other products of the mine are involved.

For example, find the number of reams of paper in 1 bale of paper. We find in the table that 2 reams = 1 bundle and 5 bundles = 1 bale. Consequently, 10 reams = 1 bale.

As another example, find the number of acres in 10 sections of land. In the table we find that 1 section equals 640 acres. Ten sections equal 10 times 640 acres or 6400 acres.

EXERCISE 1

A.

Simplify:

1. 3 ft. 29 in. =
2. 6 gal. 7 qt. =
3. 3 bu. 9 pk. =
4. 7 lb. 19 oz. =
5. 5 hr. 70 min. =
6. 4 yd. 5 ft. 14 in. =
7. 8 qt. 7 pt. 21 oz. =
8. 9 bu. 11 pk. 13 qt. =
9. 12 hr. 81 min. 170 sec. =
10. 14 qt. 9 pt. 23 oz. =

WEIGHTS AND MEASURES ──────────────────────87

Find the missing numbers:
11. 8 ft. = 7 ft. _____ in.
12. 10 bu. = 9 bu. _____ pk.
13. 5 lb. = 4 lb. _____ oz.
14. 48 min. = 47 min. _____ sec.
15. 9 qts. = 8 qts. _____ pts.
16. 8 yd. = 7 yd. _____ ft. 12 in.
17. 7 gal. 2 qt. = 6 gal. _____ qt. 2 pt.
18. 6 hr. 25 min. 35 sec. = 5 hr. 84 min _____ sec.

Add and simplify:

19. 3 ft. 5 in.
    1 ft. 3 in.
    7 ft. 2 in.
    ─────────

20. 7 ft. 9 in.
    3 ft. 8 in.
    ─────────

21. 6 yd. 3 ft. 7 in.
    3 yd. 2 ft. 8 in.
    4 yd. 4 ft. 13 in.
    ───────────────

22. 2 gal. 3 qt.
    4 gal. 2 qt.
    7 gal. 1 qt.
    ───────────

23. 7 lb. 5 oz.
    4 lb. 6 oz.
    9 lb. 7 oz.
    ──────────

24. 5 wk. 3 da.
    2 wk. 4 da.
    8 wk. 2 da.
    ──────────

25. 5 hr. 26 min.
    7 hr. 32 min.
    3 hr. 48 min.
    ────────────

26. 2 hr. 28 min. 16 sec
    5 hr. 21 min. 28 sec.
    4 hr. 37 min. 42 sec.
    ─────────────────────

Multiply and simplify:

26. 4 ft. 7 in.
         4
    ─────────

27. 5 yd. 2 ft. 6 in.
             3
    ───────────────

28. 3 ft. 9 in.
         5
    ─────────

29. 6 mi. 320 ft.
           9
    ────────────

30. 5 yd. 2 ft. 9 in.
             7
    ───────────────

B.

Multiply and simplify:

1. 4 gal. 2 qt.
   5
   ―――――

2. 7 gal. 2 pt.
   5
   ―――――

3. 9 gal. 1 qt. 2 pt.
   4
   ―――――

4. 4 bu. 1 pk.
   2
   ―――――

5. 4 lb. 9 oz.
   5
   ―――――

6. 8 min. 9 sec.
   6
   ―――――

7. 4 yr. 3 mo.
   5
   ―――――

8. 9 hr. 16 min. 8 sec.
   4
   ―――――

9. 3 yr. 6 mo.
   9
   ―――――

10. 4 hr. 9 min. 21 sec.
    6
    ―――――

11. 6 wk. 5 da. 7 hr.
    4
    ―――――

Divide and simplify:

12. 3/9 ft. 6 in.
13. 4/12 yd. 4 ft. 8 in.
14. 5/12 gal. 7 qt.
15. 6/15 qt. 6 pt.
16. 2/10 bu. 8 pk.
17. 3/11 bu. 7 pk.
18. 6/7 hr. 6 min.
19. 4/9 gal. 4 qt. 8 pt.
20. 8/31 hr. 36 min. 24 sec.
21. Change 16 feet 7 inches to inches.
22. Change 19 feet 8 inches to inches.
23. Change 9 gallons 3 quarts to pints.
24. Change 17 gallons 7 quarts to pints.
25. Change 9 pounds 5 ounces to ounces.
26. Change 21 pounds 8 ounces to ounces.
27. Change 16 bushels 2 pecks to quarts.
28. Change 43 bushels 3 pecks to quarts.

WEIGHTS AND MEASURES

29. Change 1½ gallons to cups.
30. Change 3¾ gallons to cups.

C.

1. Change 5 acres to square feet.
2. Change 9 acres to square feet.
3. Change 7911 square feet to square yards.
4. Change 86 ounces to pounds and ounces.
5. Change 359 ounces to pounds and ounces.
6. Change 529 inches to feet and inches.
7. Change 438 inches to feet and inches.
8. Change 224 pints to gallons.
9. Change 136 pints to gallons.
10. Change 646 pints to quarts.
11. Change 364 pints to quarts.
12. Change 675 cubic feet to cubic yards.
13. Change 56,376 cubic feet to cubic yards.
★ **14.** Change 2368 feet to rods and feet.
★ **15.** Change 1764 feet to rods and feet.
★ **16.** Change 163 quarts to bushels, pecks, and quarts.
★ **17.** Change 242 quarts to bushels, pecks, and quarts.
18. Change 5492 pounds to tons and pounds.
19. Change 6754 pounds to tons and pounds.
20. Add 14 quarts 1 pint, 25 quarts 1 pint, 19 quarts 1 pint, 41 quarts 1 pint, and 36 quarts 1 pint.
21. Add 75 quarts 1 pint, 27 quarts 1 pint, 35 quarts 1 pint, 66 quarts 1 pint, 58 quarts 1 pint, and 49 quarts 1 pint.
22. Add 45 pounds 6 ounces, 14 pounds 9 ounces, 88 pounds 6 ounces, 92 pounds 9 ounces, and 58 pounds 7 ounces.
23. Add 146 pounds 8 ounces, 97 pounds 8 ounces, 17 pounds 13 ounces, 47 pounds 9 ounces, and 89 pounds 15 ounces.
24. Add 7 gallons 1 pint, 11 gallons 4 pints, 9 gallons 6 pints, 5 gallons 5 pints, 27 gallons 2 pints, and 13 gallons 9 pints.
25. Add 72 gallons 3 pints, 16 gallons 7 pints, 45 gallons 5 pints, 14 gallons 4 pints, and 61 gallons 9 pints.

26. Add 7 hours 30 minutes 9 seconds, 14 hours 45 minutes 32 seconds, 11 hours 18 minutes 21 seconds, 46 hours 48 minutes, and 19 hours 49 minutes 27 seconds.

27. Add 19 hours 42 minutes 11 seconds, 42 hours 17 minutes 42 seconds, 68 hours 38 minutes 21 seconds, 37 hours 45 minutes 116 seconds, and 55 hours 22 minutes 44 seconds.

28. Subtract 18 pounds 17 ounces from 35 pounds.

29. Subtract 21 pounds 15 ounces from 64 pounds 7 ounces.

30. Subtract 46 feet 11 inches from 78 feet 9 inches.

D.

1. Subtract 37 feet 5 inches from 46 feet 2 inches.

2. Subtract 9 hours 42 minutes from 17 hours 56 minutes.

3. Subtract 15 hours 47 minutes from 19 hours 10 minutes.

4. Subtract 9 gallons 3 quarts 2 pints from 35 gallons 3 quarts 1 pint.

5. Subtract 55 gallons 3 quarts 1 pint from 65 gallons.

6. Multiply 8 feet 7 inches by 6.

7. Multiply 9 feet 11 inches by 8.

8. Multiply 17 pounds 10 ounces by 5.

9. Multiply 36 pounds 15 ounces by 9.

10. Multiply 35 hours 47 minutes by 4.

11. Multiply 65 hours 19 minutes by 7.

12. Multiply 25 feet 9 inches by 5 feet 3 inches.

13. Multiply 184 feet 9 inches by 6 feet 7 inches.

* **14.** Multiply 92 feet 14 inches by 37 feet 19 inches.

* **15.** Multiply 365 feet 21 inches by 743 feet 25 inches.

16. Divide 76 pounds 12 ounces by 4.

17. Divide 160 pounds 15 ounces by 5.

18. Divide 36 hours 50 minutes by 5.

19. Divide 160 hours 52 minutes by 8.

20. Divide 40 feet 6 inches by 2 feet 3 inches.

* **21.** Divide 74 pounds 6 ounces by 4 pounds 4 ounces.

* **22.** Find the cost of 2 pounds 9 ounces of sirloin steak at $1.26 per pound.

WEIGHTS AND MEASURES

23. In a series of hops, a plane flew for 2 hours 22 minutes, 1 hour 46 minutes, and 3 hours 43 minutes. Find the total flying time.

24. An airplane had 5 hours of fuel at takeoff. How many hours of fuel are left after the plane flies for 3 hours 36 minutes?

25. Find the total weight of a case of 2 dozen cans of pears if each can weighs 1 pound 2 ounces.

26. What is the difference in the weight of a box of soap powder weighing 3 pounds 5 ounces and the total weight of 2 smaller boxes, each weighing 1 pound 3 ounces?

27. The length of a window frame is 5 feet 8 inches and its width is 2 feet 7 inches. Find the total length of wood needed for the frame of a screen if an extra width is to be included for a center strip.

* 28. If $46\frac{1}{2}$ quarts of milk weigh 100 pounds, find the weight of 15 gallons of milk. (Round answer to nearest whole number of pounds.)

*29. Each delegate to a dry-cleaners' convention is to be provided with a badge to be made from a 6-inch piece of ribbon. At 52 cents a yard, what is the total cost of the ribbon if 1480 delegates are expected?

*30. Fourteen clerks in the White Department Store worked the following number of hours during one week: 42 hours 40 minutes, 44 hours 35 minutes, 46 hours 20 minutes, 43 hours 50 minutes, 48 hours 30 minutes, 25 hours 15 minutes, 52 hours 55 minutes, 35 hours 10 minutes, 47 hours 40 minutes, 49 hours 25 minutes, 53 hours 20 minutes, 48 hours 55 minutes, 35 hours 15 minutes, and 42 hours 45 minutes. What was the average number of hours worked by the fourteen clerks?

E.

1. Sammy sold 9 hens weighing 6 pounds 8 ounces, 7 pounds 7 ounces, 5 pounds 6 ounces, 7 pounds 9 ounces, 4 pounds 12 ounces, 5 pounds 13 ounces, 6 pounds 6 ounces, 8 pounds 4 ounces, and 6 pounds 7 ounces. Find the average weight of the hens.

2. A child weighed 7 pounds 6 ounces at birth. A year later she weighed 21 pounds 9 ounces. How much weight did the child gain in one year?

3. A fuse contains $2\frac{1}{2}$ ounces of black powder. How many fuses can be made from 750 pounds of powder?

4. How much did Sammy receive if the hens in Problem 1 brought 44 cents per pound?

5. Two pieces of wire 14 feet 8 inches and 17 feet 4 inches, respectively, are cut from a length of wire 60 feet long. How long is the remaining piece?

★6. When a car is traveling 80 miles per hour, how many feet does it go in 1 second?

7. A wheel makes 1740 revolutions per minute. How many revolutions does it make in 1 second?

★8. Change 1,000,000 seconds to weeks, days, hours, minutes, and seconds.

★9. Mr. Hilliard has employment other than his regular job. He works 3 hours 15 minutes on Monday, 2 hours 40 minutes on Tuesday, 1 hour 30 minutes on Wednesday, 4 hours 10 minutes on Thursday, and 5 hours 20 minutes on Friday. If this additional work pays him $5.50 per hour, how much extra does Mr. Hilliard earn during the week?

10. A mathematics examination consists of three parts that are to be taken in succession with no break. The first part runs for 14 minutes, the second for 23 minutes, and the third for 16 minutes. If the examination begins at 9:00 a.m., when will it end?

11. If each of 33 children at Engleton School drinks one-half pint of milk every day at school for 16 weeks, how many gallons of milk will they use?

12. How many square feet are there in 144 square yards?

13. A sanding-machine motor makes 56 revolutions per second. How many revolutions does it make per minute?

14. How many revolutions will the motor in the above problem make in three-fourths of a minute?

15. How many revolutions will the motor in Problem 13 make in two-thirds of an hour?

16. How many revolutions will the motor in Problem 13 make in 8 hours?

17. If a drill will cut through steel plate $\frac{3}{4}$ inch thick in 90 revolutions, under the same conditions how many revolutions will it take to drill through a steel plate $4\frac{1}{2}$ inches thick?

18. Ethyl gasoline is made by adding a liquid called tetraethyl

WEIGHTS AND MEASURES

lead to ordinary gasoline. One gallon of tatraethyl lead is needed for every 1200 gallons of gasoline. Three pints of tetraethyl lead will be enough for how many gallons of gasoline?

19. Have some fun discovering a person's age and month of birth. Ask the person whose age you are trying to discover to do the writing. Suppose Jack is 16 and that he was born in July. Let Jack write as follows without letting you see the computations:

Number of month of birth	7
Multiply by 2	14
Add 5	19
Multiply by 50	950
Add Jack's age, 16	966
Subtract 365, obtaining	601
Add 115	716

Jack finds the result, 716. He can now be told that his age is 16. and July, or the seventh month, is the month of his birth. The two figures on the right in the result, 716, give the age. The remaining figure or figures give the month in which the birthday comes.

20. How many squares of tile 6 inches by 6 inches would it take to cover a rectangular floor 21 feet by 15 feet?

2 | Metric system of weights and measures

The basic units in the metric system are the meter, gram, and liter. As we have already observed, the meter is the distance between two marks on a platinum bar that is kept in Paris. More specifically, it is equal to approximately 39.37 inches.

A gram is one-thousandth of a certain piece of platinum that is also kept in Paris. It takes approximately 28.35 grams to equal 1 ounce.

The liter equals 1000 cubic centimeters. Approximately 3.785 liters equal 1 gallon.

The metric system has 10 as its base, and names of the units other than meter, gram, and liter are found by adding prefixes to the names meter, gram, and liter.

PREFIXES IN METRIC SYSTEM

milli-   one-thousandth
centi-   one-hundredth
deci-   one-tenth
deka-   ten
hecto-   one hundred
kilo-   one thousand

We see that one millimeter equals one-thousandth of a meter, one centimeter equals one-hundredth of a meter, and one kilometer equals one thousand meters.

Let us consider the following tables for metric measures of length, weight, and capacity.

LENGTH MEASURE

Standard unit, 1 meter (or 1 m)
1 millimeter (mm) = 0.001 m or 0.001 meter
1 centimeter (cm) = 0.01 m or 0.01 meter
10 centimeters (cm) = 1 dm or 1 decimeter
10 decimeters (dm) = 1 m or 1 meter
10 meters (m) = 1 dkm or 1 dekameter
10 dekameters (dkm) = 1 hm or 1 hectometer
10 hectometers (hm) = 1 km or 1 kilometer
1 kilometer (km) = 1000 m or 1000 meters

WEIGHT MEASURE

Standard unit, 1 gram (or 1 g)
1 milligram (mg) = 0.001 g or 0.001 gram
1 centigram (cg) = 0.01 g or 0.01 gram
10 centigrams (cg) = 1 dg or 1 decigram
10 decigrams (dg) = 1 g or 1 gram
10 grams (g) = 1 dkg or 1 dekagram
10 dekagrams (dkg) = 1 hg or 1 hectogram
10 hectograms (hg) = 1 kg or 1 kilogram
1 kilogram (kg) = 1000 g or 1000 grams

CAPACITY MEASURE

Standard unit, 1 liter (or 1 l)
1 milliliter (ml) = 0.001 l
1 centiliter (cl) = 0.01 l
1 kiloliter (kl) = 1000 l

WEIGHTS AND MEASURES ─────────────────────────────95

It is interesting to observe that in the metric system each measure from smaller to next larger is divided by 10 and from larger to next smaller is multiplied by 10. Thus 10 mm = 1 cm; 10 cm = 1 dm, etc.

We find considerable debate concerning the relative merits of the metric and national systems of weights and measures. Only the United States and Great Britain have not adopted the metric system. Scientists everywhere use this system.

Example 1

How many grams are there in 3 kilograms? Since, from the table, 1 kilogram = 1000 grams, 3 kilograms = 3000 grams.

Example 2

Find the number of centimeters in a kilometer. We know that 100 centimeters equal 1 meter, and 1000 meters equal one kilometer. Therefore 100 × 1000 or 100,000 centimeters equal one kilometer.

Example 3

How many centiliters are there in 40 liters? We find in the table that 1 centiliter = $\frac{1}{100}$ of a liter. Therefore, 1 liter = 100 centiliters, and 40 liters = 40 × 100 or 4000 centiliters.

EXERCISE 2

1. How many centimeters are there in one meter?
2. How many millimeters are there in one meter?
3. How many millimeters are there in one centimeter?
4. How many millimeters are there in one kilometer?
5. How many centimeters are there in 400 millimeters?
6. How many millimeters are there in 60 centimeters?
7. How many grams are there in a kilogram?
8. How many milligrams are there in a kilogram?
9. How many grams are there in 4000 milligrams?
10. How many milligrams are there in 0.04 gram?
11. How many centigrams are there in 1 gram?
12. How many milligrams are there in 1 centigram?

13. How many centigrams are there in one kilogram?
14. How many kilograms are there in 200 milligrams?
15. How many centimeters are there in 200 meters?
16. How many meters are there in 350 centimeters?
17. How many milliliters are there in 1 liter?
18. How many milliliters are there in a centiliter?
19. How many centiliters are there in a kiloliter?
20. How many milliliters are there in 60 liters?
21. How many milliliters are there in 40 centiliters?
22. Convert 3,426,529 centimeters to kilometers, meters, and centimeters.
23. Subtract 16 centimeters 8 millimeters from 1 meter.
24. Find one-third of 30 meters 8 decimeters 19 centimeters.
25. Find one-fourth of 48 meters 3 decimeters 62 centimeters.
26. How many grams are there in four kilograms?
27. How many kilograms are there in 800 grams?
28. How many grams are there in 4200 milligrams?
29. How many centigrams are there in 1 gram?
30. How many centigrams are there in 60 grams?
31. How many milligrams are there in 45 grams?
32. How many grams are there in 5.25 kilograms?
33. How many kilograms are there in 400 grams?
34. How many kilometers are there in 428 meters?
35. How many centimeters are there in 5400 meters?
36. How many centiliters are there in 35 liters?
37. How far apart are the longest divisions on a meter stick?
38. What metric unit would you use in taking the dimensions (length and width) of a swimming pool?
39. What change in the position of the decimal point do we make in changing kilometers to meters?
40. What metric unit would it be best to use in measuring the distance from San Francisco to New York?
41. What change in the position of the decimal point do we make in changing from grams to kilograms?
42. Reduce 245 kilometers 573 meters to kilometers.

43. The distance from Cologne to Coblenz is 91,000 meters. Change this to kilometers.

44. Change 6574 square centimeters to square millimeters.

45. The edge of a cube is 80 decimeters long. How many centimeters is this?

46. Ice is 0.92 as heavy as water. Find the weight in kilograms of 3.5 cubic meters of ice. (A cubic meter of water weighs 1 metric ton. One metric ton weighs 1000 kilograms.)

47. A schoolroom in France is 11 meters long, 9.4 meters wide, and 4.1 meters high. How many pupils may use this room if 10.5 cubic meters of air are allowed for each pupil?

48. A line is 60 centimeters long. How many millimeters long is it?

49. What part of a meter is 90 millimeters?

50. Change 0.65 square meters to square centimeters.

51. A metric ton is equal to how many centigrams?

3 | Exponents and scientific notation

Many numbers in astronomy, chemistry, and physics are either very large or very small. Furthermore, these numbers are generally approximate numbers, with three, four, or five significant digits available. Ordinarily they are used in working out three or four digit answers. Exponents together with decimals are used to represent these quantities. We shall discuss decimals and approximate numbers in later chapters, but we would like to introduce the idea of an *exponent* at this point, and show how it is used in the representation of very large and very small dimensions and quantities.

The following examples show a few powers of 10 in terms of *exponents*.

$100,000 = 10 \times 10 \times 10 \times 10 \times 10 = 10^5$, read "ten to the fifth."

$10,000 = 10 \times 10 \times 10 \times 10 = 10^4$, read "ten to the fourth."

$1000 = 10 \times 10 \times 10 = 10^3$, read "ten to the third" or "ten cubed."

$100 = 10 \times 10 = 10^2$, read "ten squared."

$10 = 10^1$, read "ten to the first power" or simply "ten." Usually when we have a number raised to the first power we *omit* the 1.

$1 = 10^0$, read "ten to the zero power." This is a *definition*, consistent with the other powers.

$$0.1 = \frac{1}{10} = 10^{-1},$$

read "Ten to the minus one."

$$0.01 = \frac{1}{100} = \frac{1}{10 \times 10} = \frac{1}{10^2} = 10^{-2},$$

read "Ten to the minus two."

$$0.001 = \frac{1}{1000} = \frac{1}{10 \times 10 \times 10} = \frac{1}{10^3} = 10^{-3},$$

"Ten to the minus three."

$$0.0001 = \frac{1}{10,000} = \frac{1}{10 \times 10 \times 10 \times 10} = \frac{1}{10^4} = 10^{-4},$$

"Ten to the minus four."

$$0.00001 = \frac{1}{100,000} = \frac{1}{10 \times 10 \times 10 \times 10 \times 10} = \frac{1}{10^5} = 10^{-5},$$

"Ten to the minus five."

We may multiply powers of ten by adding the exponents (algebraically), and we may divide powers by subtracting exponents.

Thus:
$$10^5 \times 10^3 = 10^8$$
$$(100,000)(1000) = 100,000,000$$

And:
$$10^4 \times 10^{-3} = 10^1 = 10$$

or,
$$10,000 \times \frac{1}{1000} = 10$$

Furthermore:
$$10^3 \times 10^{-5} = 10^{-2} = 0.01$$
$$1000 \times \frac{1}{100,000} = \frac{1}{100} = 0.01$$

Consider the distance to the sun. This distance is about 93,000,000

WEIGHTS AND MEASURES _____**99**

miles. We may write this number in the following ways:

$$93{,}000{,}000 = 93 \times 10^6 = 930 \times 10^5 = 9.3 \times 10^7$$

The last of these forms using exponents, namely, 9.3×10^7, is called the *scientific notation* for the number 93,000,000. Notice that the number 9.3 in the expression 9.3×10^7, is a number between 1 and 10. The scientific notation follows the practice of writing any number as a number between 1 and 10, multiplied by the appropriate power of 10. The term scientific notation is used because this form for writing numbers is widely followed in scientific work.

Thus:
$$93{,}000{,}000 = 9.3 \times 10^7$$

The velocity of light is an important quantity in the universe. This velocity is about 186,000 miles per second. The number 186,000 may be expressed as follows:

$$186{,}000 = 186 \times 10^3 = 18.6 \times 10^4 = 1.86 \times 10^5$$

The last of these, or 1.86×10^5, is the scientific notation for 186,000. Again, 1.86 is a number between 1 and 10.

Suppose we wish to know how long it takes light to travel from the sun to the earth. The time is found by dividing the distance by the rate.

Then:

$$\text{time} = \frac{93 \times 10^6 \text{ miles}}{186 \times 10^3 \text{ miles/second}}$$

$$= \frac{93 \times 10 \times 10 \times 10 \times 10 \times 10 \times 10}{186 \times 10 \times 10 \times 10} \text{ seconds}$$

$$= \frac{93 \times 10 \times 10 \times 10}{186} \text{ seconds (dividing both numerator and denominator by } 10^3\text{).}$$

$$= \frac{1}{2} \times 10^3 \text{ seconds} = 500 \text{ seconds after dividing both numerator and denominator by 93}$$

A simpler solution would be

$$\text{time} = \frac{9.3 \times 10^7}{1.86 \times 10^5} = 5 \times 10^2 = 500 \text{ seconds}$$

This last solution is based on the scientific notation.

The preceding examples illustrate that exponents are numbers, written at the upper right of a quantity, telling how many times the quantity is used in a given product, as a factor. Thus 5^3 means $5 \times 5 \times 5$ or 125, and 4^2 means 4×4, or 16. We refer to the quantity that is used as a factor here as the *base*, and we often refer to the exponent as a power of the base. Thus in 5^3, 5 is the base, 3 is the exponent or power, and we say that 5 is raised to the third power.

A number itself, such as 7, could be written 7^1, but as a matter of convention we omit the 1.

It is necessary to agree on what we mean by "0" or zero as an exponent. By *definition*, any number (except zero) raised to the zero power, such as 7^0 or 10^0, is equal to 1. This definition is consistent with our other work in exponents. Thus we find:

$$10^3 \times 10^2 = (10 \times 10 \times 10)(10 \times 10) = 10^5$$

We *add exponents* when we multiply. And:

$$10^4 \times 10^0 = 10^{4+0} = 10^4$$

This means that when we multiply by 10^0 we get back what we had originally which is the same as multiplying by 1. That is:

$$10^4 \times (10^0) = 10^4, \text{ and } 10^4(1) = 10^4$$

Hence it is consistent to make $10^0 = 1$.

Very small numbers require *negative* exponents to express them. Thus 1 millimeter, which is commonly used in scientific measurement $= \dfrac{1}{1000}$ meter $= \dfrac{1}{10 \times 10 \times 10}$ meter $= \dfrac{1}{10^3}$ meter. By *definition* again, we say that $\dfrac{1}{10^3} = 10^{-3}$, and $\dfrac{1}{10^6} = 10^{-6}$. Decimals can also be used for these numbers. Thus $\dfrac{1}{1000} = 0.001$, hence $10^{-3} = 0.001$, and likewise $10^{-6} = 0.000001$.

Consider the radius of a hydrogen atom; it is approximately 10^{-8} centimeter, or 0.00000001 centimeter. The radius of the electron is even smaller, about 2×10^{-13} centimeter, or 0.0000000000002 centimeter. Negative exponents can be used advantageously here.

Let us use these ideas in representing numbers. Consider the following examples.

The distance to the moon is 240,000 miles; $240,000 = 2.4 \times 10^5$.

The velocity of light is 300,000,000 meters per second, or 3.0×10^8 meters per second.

WEIGHTS AND MEASURES

The number of molecules in a gram-molecular weight of a gas such as oxygen is 602,500,000,000,000,000,000,000, which is much more simply written in the form 6.025×10^{23}.

One millimeter is 0.001 meter or 1 millimeter = 10^{-3} meter.

One micron = 0.000001 meter; 1 micron = 10^{-6} meter.

The wavelength of sodium light is 0.00005893 centimeter: $0.00005893 = 5.893 \times 10^{-5}$.

The angstrom unit for measuring wavelengths of light and other radiation is 0.00000001 centimeter, or 1 Å = 10^{-8} centimeter.

EXERCISE 3

1. Change each number from exponent form to standard form.

$5 \times 10^3 = 5000$
$8 \times 10^4 =$
$1.75 \times 10^6 =$
$3.8 \times 10^8 =$
$6 \times 10^{-3} = \dfrac{6}{1000} = 0.006$
$1.97 \times 10^{-9} =$
$7.4 \times 10^{11} =$
$9 \times 10^{-5} =$
$4.6 \times 10^{-4} =$
$1.93 \times 10^{12} =$
$57 \times 10^9 =$
$1.64 \times 10^{-24} =$ (mass of hydrogen atom in grams)

2. Change each number to scientific notation.

$45,000 = 4.5 \times 10^4$
$21,000,000 =$
$93,000,000 =$
$0.00005 = 5 \times 10^{-5}$
$0.0234 =$
$0.000097 =$
$740,000 =$
$3150 =$
$0.0008 =$
$0.000056 =$
$0.000008 =$
$48,000,000,000 =$

3. The following physical constants are given in scientific notation. Write them in standard form.

2.998×10^{10} centimeters per second $= c$ (velocity of light)
6.547×10^{-27} cgs units* $= h$ (Planck's constant)
4.77×10^{-10} esu† $= e$ (charge on the electron)
1.09737×10^{5} units $= R$ (Rydberg spectra constant)
1.650×10^{-24} gram $= p$ (mass of the proton)

4. The following astronomical numbers are in standard form. Change them to scientific notation.

Diameter of Jupiter, 88,600 miles =
Mean distance from the sun to Jupiter, 483,000,000 miles =
Number of seconds in a year, 31,560,000 =
Distance to the nearest star, 25,300,000,000,000 miles =

4 | The binary system

We have already observed that our *decimal system* is written according to the base 10. Now that electronic digital computers are being rather extensively used, considerable attention is being given to the base two, which formerly appeared largely in theoretical mathematics. Two is the base of the binary system, in which all numbers are represented by the use of the two digits 1 and 0. For example, 11011 is a number written to the base 2. Let us write this number in the base 10. Then:

$$11011 = 1 \cdot 2^4 + 1 \cdot 2^3 + 0 \cdot 2^2 + 1 \cdot 2^1 + 1 \cdot 2^0$$
$$= 16 + 8 + 0 + 2 + 1$$
$$= 27 \text{ in the base } 10$$

Base 2 can be used in high-speed electronic digital computers because electrical pulses can be sent through vacuum-tube circuits at a rate of several million per second, with the presence of a charge at an instant corresponding to "1" and the absence of a charge corresponding to "0". The vacuum tubes or other electronic valves act as gates that either permit charges to pass through a circuit or

---
* The cgs system is the centimeter-gram-second system and is widely used by physicists.
† Electrostatic units.

keep them from passing through a circuit. It is not a difficult engineering problem to design circuits that will add 2 ones and carry, so that $1 + 1 = 10$, which represents 2; similarly, such circuits will add $1 + 10$ to get 11, or 3. It is a fairly difficult job in "logical analysis" or "logical design" to make computer devices that will perform general computing routines efficiently. These machines must be designed so that they will follow coded instructions. Once such machines are designed, they do exactly what they are supposed to do, without any errors, as long as the circuits are intact. They can add or subtract, multiply or divide, in a matter of a few millionths of a second.

Thus 1 in the binary system equals 1 in the decimal system.

10 in the binary system $= 1 \cdot 2^1 + 0 \cdot 1 = 2$
11 in the binary system $= 1 \cdot 2^1 + 1 \cdot 1 = 3$
100 in the binary system $= 1 \cdot 2^2 + 0 \cdot 2^1 + 0 \cdot 1 = 4$
101 in the binary system $= 1 \cdot 2^2 + 0 \cdot 2^1 + 1 \cdot 1 = 5$
110 in the binary system $= 1 \cdot 2^2 + 1 \cdot 2^1 + 0 \cdot 1 = 6$
111 in the binary system $= 1 \cdot 2^2 + 1 \cdot 2^1 + 1 \cdot 1 = 7$
1000 in the binary system $= 1 \cdot 2^3 + 0 \cdot 2^2 + 0 \cdot 2^1 + 0 \cdot 1 = 8$
1001 in the binary system $= 1 \cdot 2^3 + 0 \cdot 2^2 + 0 \cdot 2^1 + 1 \cdot 1 = 9$

We see that the sequence of natural numbers 1, 2, 3, 4, 5, 6, 7, 8, and 9, is represented in the binary system as 1, 10, 11, 100, 101, 110, 111, 1000, and 1001.

The number 100011 in the binary system is the same as 35 in the decimal system since

$$1 \cdot 2^5 + 0 \cdot 2^4 + 0 \cdot 2^3 + 0 \cdot 2^2 + 1 \cdot 2^1 + 1 \cdot 2^0 = 35$$

Let us compare this with our natural numbers or numbers to the base 10. Thus:

$$6354 = 6(1000) + 3(100) + 5(10) + 4(1)$$
$$= 6(10)^3 + 3(10)^2 + 5(10)^1 + 4(10)^0$$

It is observed that when the base is 10, the 10 is raised to various powers, whereas when the base is 2, the 2 is raised to various powers.

The most convenient method for changing from 35 in the decimal system, or base 10, to the binary system, or base 2, is by successive division.

Thus:

$$2\overline{)35}$$
$$2\overline{)17} \quad \text{remainder } 1$$
$$2\overline{)8} \quad \text{remainder } 1$$
$$2\overline{)4} \quad \text{remainder } 0$$
$$2\overline{)2} \quad \text{remainder } 0$$
$$2\overline{)1} \quad \text{remainder } 0$$
$$0 \quad \text{remainder } 1$$

Hence:

35 (base 10) = 100011 (base 2)

EXERCISE 4

Write the following natural numbers with the base 2.

1. 5
2. 17
3. 25
4. 39
5. 49
6. 54
7. 62
8. 74
9. 86
10. 98
11. 107
12. 358
13. 759
14. 2573

The following numbers are written in the binary system. Translate each into the decimal system.

15. 11
16. 110
17. 1110
18. 1111
19. 10000
20. 1101
21. 1001
22. 10011
23. 11111
24. 11010
25. 11110
26. 1110101
27. 11010111001
28. 10111110101

5 | Conversion of units

In carrying on business transactions with people in Central America, South America and continental Europe for example, we need to be able to convert from United States units to units in the metric system. We also need to know how to convert from units in

WEIGHTS AND MEASURES _____105

the metric system to United States units. By United States units we mean the units such as feet, pounds, and gallons. The following conversion table is convenient.

CONVERSION TABLE

From United States Units to Metric System Units

1 inch = 2.54 centimeters
1 square yard = 0.836 square meter
1 cubic yard = 0.765 cubic meter
1 pound = 0.454 kilogram
1 quart = 0.946 liter

From Metric System to United States System

1 meter = 39.37 inches
1 square meter = 1.196 square yards
1 cubic meter = 1.308 cubic yards
1 kilogram = 2.205 pounds
1 liter = 1.057 quarts

For example, find the number of square feet in 5 square meters. From the table just presented, 1 square meter = 1.196 square yards. Now 5 square meters = 5 × 1.196 = 5.980 square yards. But 1 square yard = 9 square feet; therefore 5.980 square yards = 9 × 5.980 square feet = 53.82 square feet. We find then that there are 53.82 square feet in 5 square meters.

As another example, find the number of centimeters in 3 yards. Three yards = 108 inches. Therefore the number of centimeters in 3 yards is 108 × 2.54 = 274 centimeters.

In the two preceding examples, as well as in the exercises that follow, we assume that the given measure is exact,* and we round the result to the appropriate number of significant figures. The number of significant figures in the answer will depend on the number of significant figures in the conversion table. Thus in the first example, 1 square meter = 1.196 square yards (4 significant figures). In the second example, 1 inch = 2.54 centimeters (3 significant figures). The answer is 274.32 centimeters, or 274 centimeters when rounded to 3 significant figures.

---

* Exact numbers and approximate numbers are treated in more detail in Chapter 7.

EXERCISE 5

A.

Change to inches:
1. 4 meters
2. 9 meters
3. 12 meters
4. 0.5 meter
5. 7.25 meters
6. 7 millimeters
7. 85 millimeters
8. 47 millimeters
9. 60 millimeters
10. 5.45 millimeters

Change to feet:
11. 7 meters
12. 52 meters
13. 300 meters
14. 1500 meters
15. 74.6 meters

Change to miles:
16. 8 kilometers
17. 42 kilometers
18. 100 kilometers
19. 365 kilometers
20. 6.4 kilometers

Change to meters:
21. 80 inches
22. 45 inches
23. $65\tfrac{1}{2}$ inches
24. $86\tfrac{3}{4}$ inches
25. 104.8 inches

Change to pounds:
26. 9 kilograms
27. 24 kilograms
28. 50 kilograms
29. 8.4 kilograms
30. 16.4 kilograms

Change to kilograms:
31. 10 pounds
32. 36 pounds
33. $8\tfrac{1}{4}$ pounds
34. 5.8 pounds
35. 200 pounds

B.
1. Change 10 meters to inches.
2. Change 25 meters to inches.
3. Change 5 miles to meters.

WEIGHTS AND MEASURES

4. Change 8 miles to meters.
5. How many meters does a boy run in the 440-yard dash?
6. How many feet are there in 40 meters?
7. How many feet are there in 55 meters?
8. Change 20 liters per second into gallons per hour.
9. Change 35 liters per second into gallons per hour.
10. Change 400 meters to miles.
11. Change 640 meters to miles.
12. Change 10 pounds per minute into grams per second.
13. Change 15 pounds per minute into grams per second.
14. How many millimeters are there in 5 feet?
15. How many millimeters are there in 8 feet?
16. A car can go 10 kilometers on 2 liters of gasoline. How many miles will it travel on a gallon of gasoline?
17. The 100-yard dash is equivalent to how many meters?
18. How many centimeters are there in an inch?
19. How many grams equal an ounce?
20. A candy bar weighs 71 grams. How many ounces is this?
21. Change 36 inches to meters.
22. How many square feet are there in a square meter?
23. Change 48 feet to meters.
24. Change 324 yards to meters.
25. Change 19 miles to meters.
26. Change 84 feet 9 inches to meters.
27. Change 47.5 meters to feet.
28. Change 6.85 kilometers to feet.
29. Change 854 centimeters to feet.
30. Change 95.6 pounds to kilograms.
31. Change 18 ounces to kilograms.
32. Change 540 grams to kilograms.
33. Change 9 pounds 6 ounces to kilograms.
34. Change 84 grams to pounds.
35. Change 6.7 kilograms to pounds.
36. Change 98 ounces to pounds.

37. Change 6.4 liters to cubic inches.

38. If air pressure is 17.6 pounds per square inch of surface, find the pressure in pounds per square centimeter.

39. The average height of a man is 5 feet 8 inches. Express this height in centimeters.

40. The barometric reading is 30.8 inches at a given time. Write this in millimeters.

41. A white rat weighed 368 grams. Find its weight in pounds.

42. A scale for a map is set up so that 1 inch = 15 miles. How many miles would be represented by 70 millimeters?

43. An admiral's gold braid is 2 inches wide. What is its width in centimeters?

44. A nautical mile is approximately 6080 feet. Convert this to meters.

45. Write in kilometers the distance from Chicago to Minneapolis. Assume that this distance is 610 miles.

46. Two airports are 148 kilometers apart. How many miles is it from one airport to the other?

47. A bag of potatoes weighs 50 pounds. Find its weight in kilograms.

48. The Empire State Building is 1248 feet high. Find its height in kilometers.

EXERCISE 6

Review

1. Add:

$$\begin{array}{r} 46.78 \\ 5.63 \\ 0.37 \\ 96.78 \\ 8.03 \\ \hline \end{array}$$

2. Divide $83\overline{)2241}$ and check.

3. Divide $128\overline{)64{,}578}$ and check.

4. Find the number of times a clock strikes in a day if it only strikes the hours.

5. Write in words $9,634,846,538.

6. Divide $64/\overline{6934}$ and check.

7. Write 679,000,000 in the scientific notation.

8. Write in words $549,683,974,506.

9. The diameter of the universe is 6×10^9 light-years. Express this in ordinary notation.

10. Write 850,000,000 in the scientific notation.

11. Tank A holds 369 gallons of oil. Tank B holds 765 gallons. Compare the capacity of the tanks by showing in decimal form the fractional part that Tank A is of Tank B.

12. A gram of radium gives off 3.7×10^{10} alpha particles per second. Write this in ordinary notation.

13. 146 is a number written in the base 10. Write this number in the base 2.

14. 685 is a number written in the base 10. Write this number in the base 2.

15. 11101101 is written in the base 2 or binary system. Write this number in the decimal system.

16. 1001000100001 is written in the base 2 or binary system. Write this number in the decimal system.

17. Express the number $8\frac{3}{4}$ as the sum of an integer and a proper fraction.

18. Multiply 86.34 by 7.65.

19. Multiply 356.8 by 48.7.

20. Divide 87.7 by 4.9.

21. Divide 876.94 by 5.78.

22. Add 38.74, 5.83, 0.69, 96.82, and 8.07.

23. Add 86.345, 921.68, 7.09, and 69.746.

24. Early morning temperatures on 5 successive days were 39°, 45°, 34°, 48°, and 30°. Find the mean or average temperature. (Add the 5 temperature readings and divide by 5.)

25. The measurements of a room were found to be 16.4 feet long, 9.8 feet wide, and 8.6 feet high. Find the area of the inner surface of the room.

26. Find the volume of the room in Problem 25.

27. The specific gravity of a substance equals the weight of a given volume of that substance divided by the weight of an equal volume of water. Find the specific gravity of glass, assuming that glass weighs 167.3 pounds per cubic foot and that a cubic foot of water weighs 62.5 pounds.

28. Find the specific gravity of ice if the ice weighs 52.7 pounds.

29. The circumference of a circle is found from the relationship $C = \pi d$, in which d is the diameter. If $\pi = 3.14159$ (approximately), find the circumference of the circle whose diameter is 6.8 inches.

30. Using the information given in Problem 29, find the circumference of the circle whose diameter is 4 miles.

31. Metal machine parts, each weighing 0.365 pound, are produced by a machine. What will 150 of these parts weigh?

32. Find the area of a painting 62.8 inches long and 38.4 inches wide.

33. There is approximately 0.45 kilogram in a pound. A man weighs 196.8 pounds. Find his weight in kilograms.

34. Simplify:

$$\frac{\frac{9}{16}}{\frac{3}{5} + \frac{5}{6}}$$

35. Simplify:

$$\frac{\frac{3}{8} + \frac{3}{10}}{\frac{4}{5} + \frac{5}{4}}$$

36. Nineteen is what fraction of 57?

37. Three-fourths is what fractional part of $\frac{27}{16}$?

38. Elsie reads 300 words a minute and Alice reads $\frac{3}{4}$ as many. How many words a minute does Alice read?

39. $84 = \frac{3}{4}$ of ?

40. 147 feet $= \frac{3}{8}$ of ? feet.

41. Last year the enrollment of a college was 4800. This year the report is that it has increased one-fifth. If this is exactly true, what is the enrollment this year?

42. A typical cost-price formula requires the cost of manufacture to be two-thirds less than the selling price. What is the selling price when the cost is $3.50?

43. Suits are on sale at two-fifths less than regular prices. If the sale price was $60, what was the regular price?

44. If the Cards have played 78 games and won 54, what is their standing? That is, what part of their games have they won? Round the answer to three decimal places.

45. Write 1970 in the Roman notation.

46. Write 1776 in the Roman notation.

47. Write MCMXLVIII in our notation.

48. Write MDCCC in our notation.

49. Ganymede, a satellite of Jupiter, is 664,000 miles from Jupiter. Write this number in the scientific notation.

50. Europa, another satellite of Jupiter, is 415,000 miles from Jupiter. Write this number in the scientific notation.

51. The force of gravity between the sun and the earth written in dynes is 324×10^{25}. Write this in the ordinary notation.

52. The sun's mass equals 199×10^{31} grams. Write this in the ordinary notation.

53. Write 1000100011111, which is according to base 2, as it would appear to the base 10.

54. Write 1111000011110101, which is according to base 2, as it would appear to the base 10.

55. 689 is a number in base 10. Write it according to base 2.

56. 1244 is a number in base 10. Write it according to base 2.

Chapter 6

Percentage

1 | Definitions

Percentage is widely used by the modern businessman. Merchants use it in computing markups and discounts, and banks use it in lending and borrowing money. Percentage is also utilized by local, state, and federal agencies in collecting taxes; by insurance companies in determining rates; and by newspapers and magazines in describing such things as business and population changes, accident rates on the highways, and agricultural production. All of us come in contact with it to such an extent that it is very important that we understand clearly what it is and how it is used.

The term *per cent* comes from the Latin *per centum*, which means *by the hundred*. Thus, the original meaning of 6 per cent was 6 out of every hundred. The per cent sign (%) came into being in the fifteenth century but was not generally used until many years later. It is interesting that the per cent sign contains two zeros, one for each zero in 100.

The meaning of per cent is now much broader than the original concept of by the hundred. We not only use 6% and 60%, but we also deal with 600% and sometimes even 6000%. According to modern usage, a *per cent* is a fraction having 100 for the denominator.

Percentage is that portion of arithmetic dealing with *per cent*. However, the terms *per cent* and *percentage* are frequently used interchangeably. We shall discuss the meaning of *percentage* further as we proceed in this chapter.

2 | Reduction of a per cent to a decimal fraction

Since per cent is only another way of saying hundredths, $5\% = 5$ hundredths $= \frac{5}{100} = 0.05$. *A per cent can be reduced to a decimal fraction by dropping the per cent sign and moving the decimal point two places to the left.* For example:

$$6\tfrac{1}{2}\% = 0.06\tfrac{1}{2}, \qquad 350\% = 3.50, \quad \text{and} \quad 5000\% = 50.00$$

EXERCISE 1

Change the following per cents to decimal fractions.

1. 3%
2. 6%
3. 5%
4. 8%
5. $4\tfrac{1}{2}\%$
6. $9\tfrac{1}{2}\%$
7. 62%
8. 95%
9. 74%
10. 53%
11. 145%
12. 328%
13. 640%
14. 775%
15. 3260%
16. 7580%
17. 2873%
18. 5965%
19. 39.73%
20. 68.29%
21. 637.5%
22. 794.8%
23. 153.7%
24. 536.4%
25. $4039\tfrac{5}{8}\%$
26. $11{,}057\tfrac{2}{3}\%$
27. 3469.75%
28. 57,039.125%

29. A roast is described as being 17% bone. Write this per cent as a decimal fraction.

30. A student on the business staff of a college paper will receive a commission of $12\tfrac{1}{2}\%$ on all the advertising space he sells. Write his commission rate as a decimal fraction.

31. The rental on the Blake Apartment is to be increased 7.5%. Write this percent as a decimal fraction.

32. $53\tfrac{1}{2}\%$ of a college class in mathematics commutes. Write this as a decimal fraction.

3 | Reduction of a decimal fraction to a per cent

We may write the decimal fraction 0.03 in its equivalent form $\frac{3}{100}$ or 3%, since, as we have recently pointed out, a per cent is a

PERCENTAGE _____115

fraction having 100 for the denominator. A convenient rule to follow in changing a decimal fraction to a per cent is *to move the decimal point two places to the right and annex the per cent sign.*

Other examples are $0.07 = 7\%$, $0.0675 = 6.75\%$, and $0.04\frac{1}{2} = 4\frac{1}{2}\%$. Consider $0.3\frac{1}{2}$. First write it as 0.35. Changing to per cent, this becomes 35%.

EXERCISE 2

Change the following decimal fractions to per cents.
1. 0.03
2. 0.05
3. 0.0625
4. 0.0525
5. 0.0833
6. 0.0775
7. $0.06\frac{1}{4}$
8. $0.2\frac{1}{2}$
9. $0.07\frac{3}{4}$
10. $0.03\frac{5}{8}$
11. 0.6873
12. 0.7936
13. 0.3743
14. 0.5638
15. 7.654
16. 8.493
17. 86.93
18. 48.65
19. $864.37\frac{1}{4}$
20. $729.48\frac{1}{4}$
21. 0.06857
22. 0.05497
23. $49.32\frac{1}{2}$
24. $86.49\frac{3}{4}$
25. 692.37
26. 746.93

27. Eighteen-carat gold is 0.75 parts gold, while copper and silver combine to make up the remaining 0.25 parts. Change both of these decimal fractions to per cents.

28. Some kinds of fish lose 0.29 of their original weight in cleaning. Change this decimal fraction to a per cent.

29. The single discount rate, which is the equivalent of two given successive discount rates, is found by multiplying the two given discount rates and subtracting this product from the sum of the two given discount rates. Find the single discount rate that is equivalent to two successive discount rates of 12% and 15%. Express the answer in per cent.*

30. From the sum of 58.7%, 0.0463, 5.76, and 48.69% subtract the sum of 0.1962, 7.46%, $43\frac{1}{4}\%$, and 0.0576. Express the result in per cent.

---

* Discounts reduce the price of goods or the amount of a bill or debt and are frequently given to encourage payment before the due date.

4 | Reduction of a per cent to a common fraction

As we have noticed, per cent is another way of saying hundredths. Thus, $5\% = 5$ hundredths $= \frac{5}{100} = \frac{1}{20}$. The equivalent common fraction to 5% is $\frac{1}{20}$. Other examples are:

$$8\% = \frac{8}{100} = \frac{2}{25}; \qquad 12\frac{1}{2}\% = \frac{12\frac{1}{2}}{100} = \frac{\frac{25}{2}}{100} = \frac{25}{200} = \frac{1}{8};$$

and

$$62\frac{1}{2}\% = \frac{62\frac{1}{2}}{100} = \frac{125}{200} = \frac{5}{8}$$

These common fractions should always be reduced to lowest terms.

The following per cents reduce to the so-called business fractions that are very commonly used in the business world.

$$50\% = \tfrac{1}{2} \qquad\qquad 20\% = \tfrac{1}{5}$$
$$33\tfrac{1}{3}\% = \tfrac{1}{3} \qquad\qquad 40\% = \tfrac{2}{5}$$
$$66\tfrac{2}{3}\% = \tfrac{2}{3} \qquad\qquad 60\% = \tfrac{3}{5}$$
$$25\% = \tfrac{1}{4} \qquad\qquad 80\% = \tfrac{4}{5}$$
$$75\% = \tfrac{3}{4} \qquad\qquad 16\tfrac{2}{3}\% = \tfrac{1}{6}$$
$$83\tfrac{1}{3}\% = \tfrac{5}{6} \qquad\qquad 8\tfrac{1}{3}\% = \tfrac{1}{12}$$
$$12\tfrac{1}{2}\% = \tfrac{1}{8} \qquad\qquad 41\tfrac{2}{3}\% = \tfrac{5}{12}$$
$$37\tfrac{1}{2}\% = \tfrac{3}{8} \qquad\qquad 6\tfrac{1}{4}\% = \tfrac{1}{16}$$
$$62\tfrac{1}{2}\% = \tfrac{5}{8} \qquad\qquad 5\% = \tfrac{1}{20}$$
$$87\tfrac{1}{2}\% = \tfrac{7}{8} \qquad\qquad 4\% = \tfrac{1}{25}$$

EXERCISE 3

Change the following per cents to common fractions reduced to lowest terms.

1. 6%
2. 8%
3. 4%
4. 9%
5. 40%
6. 60%
7. 175%
8. 110%
9. $5\tfrac{1}{2}\%$
10. $7\tfrac{3}{4}\%$
11. $4\tfrac{1}{4}\%$
12. $6\tfrac{3}{8}\%$
13. 46.5%
14. 81.25%
15. 35.75%

16. 46.8%   **17.** $124\frac{1}{2}$%   **18.** $65\frac{3}{4}$%
19. $150\frac{1}{4}$%   **20.** 25.5%   **21.** 0.6345%
22. 0.2546%

23. Milk contains 4.4% butterfat. Write this per cent as a common fraction.

24. Axley paid 60% of the price of an automobile as a down payment. Write this per cent as a common fraction.

25. Instead of multiplying 144 by $12\frac{1}{2}$%, what fractional part of 144 should we take? Find $12\frac{1}{2}$% of 144.

26. Express $\frac{45}{100}$, 45%, and 0.45 as twentieths.

5 | Reduction of a common fraction to a per cent

We have already observed that a common fraction is transformed into a decimal fraction by dividing the denominator into the numerator. *The decimal fraction is changed to a per cent by moving the decimal point two places to the right and annexing the per cent sign.* Thus $\frac{3}{5} = 0.60 = 60$%, and $\frac{9}{2} = 4.50 = 450$%.

EXERCISE 4

Change the following common fractions to per cents.

1. $\frac{2}{5}$ **2.** $\frac{1}{4}$ **3.** $\frac{3}{8}$

4. $\frac{5}{8}$ **5.** $\frac{15}{5}$ **6.** $\frac{17}{4}$

7. $\frac{24}{3}$ **8.** $\frac{50}{2}$ **9.** $\frac{15}{8}$

10. $\frac{19}{5}$ **11.** $\frac{500}{400}$ **12.** $\frac{700}{400}$

13. $\frac{750}{125}$ **14.** $\frac{742}{106}$ **15.** $\frac{2000}{5}$

16. $\frac{1200}{6}$ **17.** $\frac{71}{4}$ **18.** $\frac{92}{5}$

In Problems 19 and 20, carry the division to four decimal places, round to three, and then change to per cent.

19. $\dfrac{43}{86}$ **20.** $\dfrac{236}{47}$

In the following four problems, arrange in order of size, beginning with the smallest (Problem 21 is worked for you):

21. $\dfrac{4}{3}$, 125%, $\dfrac{3}{4}$, $1\tfrac{1}{8}$

Solution:

$$\dfrac{4}{3}, 1\tfrac{1}{4}, \dfrac{3}{4}, 1\tfrac{1}{8}$$

$$\dfrac{4}{3}, \dfrac{5}{4}, \dfrac{3}{4}, \dfrac{9}{8}$$

Rewriting the fractions with the common denominator 24, we have:

$$\dfrac{32}{24}, \dfrac{30}{24}, \dfrac{18}{24}, \dfrac{27}{24}$$

Rearranging the fractions from smallest to largest:

$$\dfrac{18}{24}, \dfrac{27}{24}, \dfrac{30}{24}, \dfrac{32}{24}$$

or

$$\dfrac{3}{4}, \dfrac{9}{8}, \dfrac{5}{4}, \dfrac{4}{3}$$

or

$$\dfrac{3}{4}, 1\tfrac{1}{8}, 1\tfrac{1}{4}, \dfrac{4}{3}$$

or

$$\boxed{\dfrac{3}{4}, 1\tfrac{1}{8}, 125\%, \dfrac{4}{3}}$$

is the answer.

PERCENTAGE   119

Another solution consists of changing into decimal form and rearranging according to size.

$$\frac{4}{3}, 125\%, \frac{3}{4}, 1\tfrac{1}{8}$$

| 1.33, | 1.25, | 0.75, | 1.125 |
| 0.75, | 1.125, | 1.25, | 1.33 |

$$\boxed{\frac{3}{4}, 1\tfrac{1}{8}, 125\%, 1.33}$$

is the answer.

22. $58\%, \dfrac{18}{23}, \dfrac{3}{5}, 0.6\tfrac{3}{5}$ **23.** $\dfrac{1}{10}, 0.25\%, \dfrac{1}{75}, \dfrac{19}{24}$

24. $62\tfrac{1}{2}\%, \dfrac{3}{8}, \dfrac{3}{4}, 0.4\tfrac{3}{4}$ **25.** $46.5\%, \dfrac{2}{5}, 0.4\tfrac{3}{5}, \dfrac{17}{16}$

26. Read the following numbers and tell which of them are equal: 4; 400%; $\tfrac{400}{100}$; 7; 700%; $\tfrac{240}{60}$.

27. The Gamma Mu team lost 8 games out of 48 games played. The games lost were what part of the games played? Change this common fraction to a per cent.

28. A team won ⅜ of the games it played. What per cent did it win?

6 | The three cases of percentage

What is 25% of 64? To determine this the per cent is first reduced to a decimal fraction, then 25% of 64 = 0.25 of 64 = 0.25 × 64 = 16. In the latter multiplication, the multiplier, 0.25, is called the *rate*, the multiplicand, 64, is the *base*, and the product, 16, is the *percentage*. These three quantities are connected by the relationship

$$Rate \times Base = Percentage$$

If any two of the three quantities are given, the third can be found. Consequently, there are three basic cases of percentage: (1) given the rate and the base, to find the percentage; (2) given the base and the percentage, to find the rate; (3) given the rate and the percentage, to find the base. Let us consider each of these cases.

1. *Find the percentage.* The problem of finding a per cent of a number is another way of finding a fractional part of the number. In finding a per cent of a given quantity, first reduce the per cent to either a common fraction or a decimal fraction, and multiply.

For example, find $12\frac{1}{2}\%$ of 64.08. Reducing to a common fraction, $12\frac{1}{2}\%$ of $64.08 = \frac{1}{8}$ of $64.08 = 8.01$. Reducing to a decimal fraction, $12\frac{1}{2}\%$ of $64.08 = 0.125 \times 64.08 = 8.01$. Here, 0.125 is the rate, 64.08 is the base, and 8.01 is the percentage.

2. *Find the rate.* A rather common problem is to determine the per cent one number is of another. Thus, what per cent of 96 is 24? Here, the base, 96, and the percentage, 24, are given. Now, by the fundamental relationship just given, rate $\times\ 96 = 24$. This means that the product of two numbers is 24, and that one of the two numbers is 96. Division is defined as the process of finding either of two numbers when their product and the other number are known. From the definition of division, the other number may be found by dividing; then, rate $= \frac{24}{96} = 0.25 = 25\%$.

As another example, find what per cent of 144 is 54. Now, rate $\times\ 144 = 54$, or rate $= \frac{54}{144} = 0.375 = 37.5\%$.

3. *Find the base.* To find the base when the rate and percentage are given, reduce the rate either to a common fraction or to a decimal fraction.

For example, a house rents for $1800 per year and the rental represents 10% of the value of the house. Find the value of the house. Now, $0.10 \times$ base $= \$1800$. Here the product of two numbers and one of the numbers are given. From the definition of division, the other number can be found by dividing. Now, base $= \frac{\$1800}{0.10} = \$18{,}000$.

Sometimes it is confusing in percentage questions to determine which type of problem is involved. Try to be careful not to confuse the base and the percentage. Sometimes the percentage or rate may be stated indirectly, and it is necessary to read the problem very carefully.

EXERCISE
5

In any of these exercises if the answers do not come out even they should be rounded to two decimal places.

PERCENTAGE

1. Find 14% of 85.
2. Find 76% of 95.
3. What is 65% of 43?
4. What is 42% of 16?
5. Find 98.6% of 142.
6. Find 47.8% of 673.
7. 45 is what per cent of 846?
8. 22 is what per cent of 78?
9. 674 is what per cent of 1865?
10. 172 is what per cent of 684?
11. Five hundred fifty-three and forty-seven hundredths is what per cent of eight thousand three hundred sixty-six and three hundred forty-eight thousandths?
12. Two hundred eighty-four and sixty-nine hundredths is what per cent of four hundred sixty-nine and sixty-eight hundredths?
13. Find a number such that 42 per cent of it is 79.
14. Find a number such that 17 per cent of it is 68.
15. Find a number such that 62 per cent of it is 98.
16. Find a number such that 22 per cent of it is 146.
17. Forty-six and eighty-seven hundredths is 56 per cent of what number?
18. Two hundred thirty-eight and eighty-five hundredths is 84 per cent of what number?
19. Find a number such that 48 per cent of it is 647.8.
20. Find a number such that 74 per cent of it is 274.8.
21. Find a number such that $27\frac{1}{4}$ per cent of it is 849.6.
★ 22. Find a number such that $52\frac{3}{8}$ per cent of it is 247.6.
23. 355 is what per cent of 926.8?
★ 24. 6543 is what per cent of 14,580?
25. 746 is what per cent of 452?
26. 259 is what per cent of 678?
27. 864 is what per cent of 351?
28. 564 is what per cent of 765?
★ 29. 3692 is what per cent of 643.9?
★ 30. 7358 is what per cent of 648.9?
31. Find 143% of 672.
32. Find 185% of 729.
33. Find 678% of 259.
34. Find 859% of 347.
35. Find 354.7% of 854.
36. Find 46.75% of 428.
37. Find a number such that 56% of it is 69.

38. Find a number such that 83% of it is 693.

★ **39.** Find a number such that 546.3% of it is 758.

★ **40.** Find a number such that 3287% of it is 539.

41. If 74 per cent of the total weight of a hog can be made into edible products, how many pounds of food can be obtained from a 204-pound hog?

42. If 28 per cent of a 1980-pound load of peaches was spoiled, how many pounds of peaches were spoiled?

43. An automobile listed at $3250 was sold at a 15% discount. What was the selling price?

44. Mack's Drive-In buys a box of 24 candy bars for 80 cents and sells them for 5 cents each. What per cent of the sales is profit?

45. Mr. Randall, with a salary of $7200 a year, was given a 10% cut. Later his salary was increased 10%. What was his salary after the increase?

46. A rug listed at $124 was sold during a sale for $88. What was the per cent of discount?

47. A house rents for $1500 per year. If the rent represents 12% of the value of the house, what is the house worth?

48. James is paid a salary of $4000 a year plus a commission of 4% on his sales. What must be the amount of his sales during the year if he wishes to receive a total salary of $10,000?

49. The list price of a mathematics textbook is $7.50, less a discount of 15% to teachers. What is the net price of the book to a teacher?

50. Jackson Brothers failed in business. Their assets amounted to $43,895 and their liabilities to $124,630. What per cent of their liabilities did they pay? How much did a creditor receive who was owed $1850?

51. Assuming that a fish loses 28% of its weight in cleaning, how many pounds of cleaned fish can be obtained from 1325 pounds of raw fish?

52. An apartment is insured for three-fourths of its assessed value of $85,000. If the insurance rate is $\frac{1}{4}$% a year, what is the premium per year?

PERCENTAGE _____123

53. A concrete mixture is 1 part cement, 2 parts sand, and 2 parts gravel by weight. What per cent of the mixture is cement? Sand? Gravel?

54. In the above problem, how many pounds of cement are in 8 tons of the mixture?

55. According to the data in Problem 53, how many pounds of the mixture can be made with 45,000 pounds of gravel?

56. Green gold is 60% gold, 35% silver, and 5% copper. How many ounces each of gold, silver, and copper are in 4800 ounces of green gold?

57. A 4200-pound automobile contains 85 pounds of chromium, 120 pounds of lead, and 462 pounds of rubber. What per cent of its total weight is chromium? Lead? Rubber?

58. Five tons of a commercial feed contains 30 pounds of salt. What per cent of the feed is salt?

59. Duralumin is one-half of 1% magnesium. How many pounds of duralumin can be made with 420 pounds of magnesium?

60. Wrought iron is 99.8 per cent pure iron. How many pounds of impurities are in 356.74 tons of wrought iron?

61. A. W. Fisher invested $12,800 and lost $460. What per cent of his investment did he lose?

62. An aluminum alloy called lynite is used for piston rods in motors. Lynite includes 11% by weight of copper. How many pounds of lynite would contain a total of 960 pounds of copper?

63. Some steel contains three-fourths of 1% of carbon. How many tons of this steel can be made with 6400 pounds of carbon?

64. A flock of 120,000 hens on the Early Bird Poultry Farm averages 9900 dozen eggs per day. What per cent production is this?

65. A hard silver solder, used for soldering metals where the job must withstand very severe strains, is an alloy including 6.8% by weight of zinc. How many pounds of this alloy will be made by adding 15.9 ounces of zinc to the proper proportions of the other materials?

7 | Arithmetic in the business world

Let us now consider some basic topics in business arithmetic.

If a storekeeper sells an article for $5 and this article *cost* him $3, we say that he has made a *gross profit* of $2 on the sale. If he decides to sell one of these articles for $1.50, he has had a *loss* of $1.50 on the sale. *Gross profit* is the difference between the *selling price* and the *cost*. *Gross profit* is rarely ever the actual profit made by the merchant. Such expenses as advertising, payrolls, rent, taxes, etc., are not included when the *gross profit* is computed. *Gross profit* is also called markup. *Net profit* is the *actual profit* left over after *all expenses* have been deducted from the *gross profit*. *Gross margin* is another term for *gross profit*. Thus if net sales for a day are $4683 and the cost of goods sold is $2569, then the *gross margin* or *gross profit* is $4683 − $2569 = $2114.

One of the important things a businessman needs to understand about his business is the cost of selling his product or his *selling expense*. *Selling expense* includes commissions, salesmen's salaries, etc. The businessman needs to know the selling expense per dollar of sales or the ratio of selling expense to sales. For example he may compare the months of March and August. If there is a significant difference between the two months, he might decide to change the size of the sales force.

The *selling expense* per dollar of sales is found by dividing the total sales expense by the total sales. If the total sales for a certain day are $312 and the selling expenses for that day are $40, then the selling expense per dollar of sales is 40 ÷ 312 = 0.1282. It costs $0.1282 to sell $1 worth of goods. This result of course needs to be small if the man is to continue his business. In order to compare selling expenses it is frequently necessary to carry this division to several places.

Dollar markup is another frequently used business term. It may be the *per cent markup times the cost*, or it may be the *per cent markup times the selling price*, depending on which is used as the base. Per cent markup times the cost is also known as *profit on cost*, whereas per cent markup times the selling price is referred to as *profit on selling price* or *profit on sales*. Regardless how the dollar markup is arrived at, we always have the basic relationship, "selling price equals cost plus dollar markup."

PERCENTAGE ———————————————————————125

In the paragraph just above we have used the phrase *profit on cost*. Of course this means that the selling price is arrived at by taking a per cent of the cost. For example, a television set costs a dealer $86.50. At what price should he sell it to have a markup of 40% on the cost? The solution is as follows:

Cost	$86.50
Profit or markup = 40% or 0.40 × 86.50	34.60
Selling price	$121.10

We have also referred to *profit on selling price*. Although dollar markup can be conveniently calculated when profit is figured on cost, merchants usually make their markups a per cent of the selling price, rather than a per cent of the cost. They are of course using the selling price rather than the cost as a base. Probably the main reason for this procedure is that the sales records tell the merchant much more easily what his sales for the day, week, or month have been than his records can tell him what the costs of certain goods sold were. This information is easily available from the cash register. His other records are generally in reference to his sales. Advertising, rent, salaries, salesmen's commissions, taxes, utilities, and so forth are usually posted as costing a certain per cent of his sales. Since almost every expense is referred to sales it is practical to use the sales figure as the base for the markup.

As an example, a furniture store has an electric refrigerator priced to sell at $620. If the store figures its profit at 45% of the sale price, find the profit.

Solution: 45% of $620 or 0.45 × 620 = $279 profit. This represents gross profit of course and does not include freight, store rent, clerical salaries and so forth.

Consider the example of a hat that cost $10.00 being sold at a markup of 35% on the cost. Find the per cent markup on the selling price.

The solution is as follows:

Cost =	$10.00
Markup = 35% or .35 × $10.00 or	3.50
Selling price =	$13.50
Per cent markup on selling price =	

$$\frac{\$3.50}{\$13.50} \text{ or } 0.259 \text{ or } 25.9\%$$

EXERCISE 6

1. An article costing $15.00 is sold at a markup of 30% on the cost. Find the per cent markup on the selling price.

2. Lay's store wants to buy some shirts to sell at $5.50. It wishes to show a gross profit of 43% on the selling price. How much may it pay for the shirts?

3. The Soft Wear Shoe Store wants to buy some shoes to sell at $26.50. It wishes to show a gross profit of 45% on the selling price. How much may it pay for the shoes?

4. An article costing $24.00 is sold at a markup of 30% on the cost. Find the per cent markup on the selling price.

5. All prices at a fire sale are to be reduced by one-fourth. What should be the sale price on an article formerly priced at $18.80?

6. Gloves were bought at $48.00 a dozen pair and sold at $5.50 per pair. What was the per cent profit on cost, and what was the per cent profit on selling price?

7. A dress is advertised for $30, which represents a 30% reduction from its former price. What was the former price?

8. How much must a salesman, working on a 12% commission, sell per month to earn $860 a month?

9. A gas bill of $18.54 may be paid in full within ten days for $17.14. What per cent saving is involved?

10. Membership in a consumers' union entitles a member to a 16% discount on all purchases in certain stores. What will some furniture listed at $398.50 cost a member if he lives in a state where there is a 5% sales tax?

11. Gayle's Music Store has a saxophone priced at $192.00. This price represents a markup of 45% on cost. Find the cost.

12. An employee earning $140 a week was given an 18% increase. A short time later business conditions necessitated an 18% decrease for all employees. What was the employee's salary after the decrease?

13. A drugstore sells cosmetics at a markup of 35% on the selling price. Find the selling price for a bottle of lotion on which the markup was $1.90.

14. In Problem 13, find the cost, and the cost as a per cent of the selling price.

15. A buyer purchased a job lot of 350 linen table cloths for $2642. If he sold 35 of the cloths at $13.10 each, 105 at $11.80 each, 120 at $10.45 each, and the remainder at $9.75 each, find the total selling price.

16. In Problem 15, find the total markup, and the markup per cent on the selling price.

17. Rawl's Toyland sells toys at a markup of 38% on cost. Find the cost for a toy on which the markup is $3.40.

18. In Problem 17, find the selling price, and the selling price as a per cent of the cost.

19. Black Wholesale Plumbing prices a faucet at $17.40. If the cost was $14.50, find the markup, the markup as a per cent of selling price, and the cost as a per cent of the selling price.

20. The Center, to celebrate its twenty-fifth anniversary, offers a 25% reduction. If living room furniture was originally priced at $379, what is the reduced price? If a 4% sales tax is added to the new price, what is the total cost to a purchaser?

21. A suit was priced at $95. On a sale, the price was reduced by one-fifth. After the sale, the sale price was increased by one-fifth. What was the latest price?

22. A store paid $28 for table radios. If the markup was 32 per cent of the selling price, find the selling price.

23. Clem's Sport Center sold golf balls at a markup of $2.60 per dozen. If the cost was 60% of the selling price, find the selling price per dozen.

24. Charles and Sons allow discounts of $12\frac{1}{2}$% and 10%, and an additional discount of 2% if the bill is paid within 10 days. Find the net cost of a bill of $346.50 if paid within 10 days.

25. A buyer for a store put a markup of 55% of the selling price on some women's coats that he sold at $67.50. Find the cost of the coats.

26. Will a buyer be better off to take discounts of 5% and 10%, or discounts of 10% and 5%?

27. In computing discounts, a new clerk allowed a discount of 25% when he should have allowed discounts of 15% and 10%. What was the amount of error on a bill of $1650?

28. An article is sold for $45 and $12\frac{1}{2}\%$ is lost. What should it be sold for to gain $12\frac{1}{2}\%$?

Solution:

Let 100% = the cost price.

Then

$$87\frac{1}{2}\% = \text{selling price by the first condition}$$

and

$$112\frac{1}{2}\% = \text{selling price by the second condition}$$

It is thus seen that the two selling prices are to each other as 7 to 9 and that the second is nine-sevenths of the first.

29. Malone Department Store uses $120,000 capital. It has $25,000 expenses and is expected to show a profit of 8% of the capital and expenses. What amount of profit is expected?

30. Graham Auto Supply sells side-view mirrors at a selling price that is 125% of the cost. If the markup was $1.90, find the selling price.

31. Five Points Furniture Store paid $289.95 for some color television sets. If the markup was 42 per cent of the selling price, find the selling price.

32. Ashley Men's Apparel buys suits at $78.50 and the markup is to be 45 per cent of the cost. Find the retail price they will charge for each suit.

33. Will a buyer be better off to take discounts of 12% and 8%, or discounts of 8% and 12%?

34. A buyer for a store put a markup of 48% of the selling price on some ladies' dresses that he sold at $27.95. Find the cost of the dresses.

35. Towson Center allows discounts of 5% and 10%, and an additional discount of 3% if the bill is paid within 10 days. Find the net cost of a bill of $429.87 if paid within 10 days.

Chapter 7

Approximate Numbers

1 | Measurements and approximate numbers

People frequently have the impression that all numbers are exact and that computations involving such numbers, if performed without error, result in exact numbers. Such a situation is not always true, of course.

Exact numbers and approximate numbers form an interesting topic for discussion.

Counting is done frequently by all of us in our daily living. *When units that cannot be divided are counted and described as a certain number, that number is exact.* This means it is 100 per cent perfect. If the attendance at a basketball game is reported as 4362 on a certain night, it means exactly 4362 people. If a merchant counts and finds that he has 42 dresses and 25 hats, he means just that. If you look and find that you have $32 in your billfold, you mean exactly that amount. Calculations with exact numbers, when correctly performed, produce exact results. If 40,000 people buy football tickets at $5.00 each, the amount is $200,000.00. This $200,000.00 is the result of counting an indivisible unit, the dollar, and is an exact number.*

Numbers resulting from counting indivisible units are considered

---

* We speak of the dollar as being indivisible because the silver dollar and the dollar bill, as such, cannot be subdivided.

to be exact. Numbers representing measurement are considered approximate numbers. We should not think of an approximate number as one that has been carelessly obtained. An approximate number is a number that does not express absolute precision, but it records to a certain degree of accuracy. We should not look upon computation with approximate numbers as computation that has not been carried out carefully. There never has been such a thing as an exact measurement of weight, time, length, area, volume, temperature and so forth. Many attempts have been made to obtain absolute standards of measurements, but no perfect measurements have ever been made.

All numbers that represent measurements, as well as many other numbers, are approximate numbers. Practically all the numbers taken from handbook tables are approximate. There are many such tables, some of which contain thousands of approximate numbers.

In discussing approximate numbers, the terms "units of measurement," and "significant digits" need to be understood.

2 | Units of measurement

The unit of measurement is the smallest unit used in carrying out a measurement. If a quart jar were used to measure the water placed in a washtub, the unit of measurement would be the quart jar. If the hand were used to measure the height of a horse, the hand would be the unit of measure. If an unmarked board one yard long were used to measure the length of a fence, the yard would be the unit of measurement. In working with approximate numbers we shall want to look for the unit of measurement that is used. In case the unit of measurement used is not obvious, it is best to use the least precise unit suggested by the given data.

In describing the distance from the earth to the sun as 93,000,000 miles it is not clear whether the distance has been measured to the nearest mile, the nearest thousand miles, or the nearest million miles. The least precise unit of measurement in this distance is 1,000,000 miles. Consequently this is the unit we shall use.

Consider the width of a desk given as 3.6 feet. Here the unit of measure is readily seen to be one tenth of a foot. The unit of measurement of 0.1 foot means that the edge of the desk has a length of 3.6 ± 0.05 feet. The measurement is recorded as 3.6 feet, but the real measurement lies anywhere between 3.55 and 3.65 feet. If the measurement of the width of the desk is listed as 3.64 feet, then 0.01 foot is immediately suggested as being the unit of measurement. The width of the desk is 3.64 ± 0.005 feet, and the true measurement is between 3.635 and 3.645 feet.

Another example is a tank said to contain 6743 gallons. Here the gallon is evidently the unit of measure. The tank would contain 6743 ± 0.5 gallons and the true measure is between 6742.5 and 6743.5 gallons.

If a vial contains 3.5679 ounces, the unit of measurement is 0.0001 ounces. The content should be expressed as 3.5679 ± 0.00005 ounces. Here the measurement is anywhere between 3.56785 and 3.56795 ounces.

If the length of a book is called $8\frac{1}{8}$ inches, the unit of measurement is $\frac{1}{8}$ inch, and the true measurement is between $8\frac{1}{16}$ inches and $8\frac{3}{16}$ inches.

3 | Significant figures

It is important to understand what is meant by the expression "significant digits," since the counting of these digits gives us one of the best ways of describing the accuracy of approximate numbers. In approximate numbers, if all the digits except the last one in a number are correct, and if the error in the last digit is not greater than one-half the unit of measurement used, then all of the digits in the approximate number are significant. Zeros may or may not be significant. Thus in a distance of 25 feet, with a unit of measurement of 1 foot, the 2 and the 5 are said to be significant. In a distance of 250 feet, with a unit of measurement of 1 foot, the 2, 5, and 0 are counters and consequently are significant. If the distance 250 feet is recorded and the unit of measurement is 10 feet, then the 25 means that there are 25 of the given units in the total distance of 250 feet. The 0 to the right of the 25 is not significant but it needs to be present as a place marker.

If a storage tank is said to contain 3600 gallons and the unit of measure is understood to be 1 gallon, the 3, 6, and the two zeros are counters and consequently every figure in the 3600 is significant. Again taking the figure 3600 gallons, but this time using the unit of measure as 100 gallons, we see that there are 36 of the 100 gallon units in the 3600 gallons. The 3 and 6 are significant but the zeros are not. However, they need to be present as place holders.

The following are the cases where zeros are significant:

1. They are significant when they fall between digits that are significant. Thus in 603.04 feet, the unit of measurement is 0.01 foot, there is a total of 60,304 units and all 5 of the digits 6, 0, 3, 0, and 4 are significant. Also in 3004.2 pounds, the unit of measurement is 0.1 of a pound. There is a total of 30,042 units and all 5 of the digits 3, 0, 0, 4, and 2 are significant.

2. Zeros are significant when they stand at the right of a significant digit in a decimal fraction. For example, in 0.0070 inches the unit of measure that is implied is 0.0001 inch. Here there is a total of 70 units, and the significant digits are the last two on the extreme right, namely the 7 and the 0.

3. Zeros are significant when they stand after the decimal point in a mixed number. Consider the measurement 125.0 gallons. The unit of measurement here is apparently 0.1 of a gallon, the total number of units is 1250, and the significant digits are 1, 2, 5, and 0. In the number 603.04 referred to in a preceding paragraph, we have another example of a zero coming after the decimal point in a mixed number. Here as we have pointed out earlier we have 5 significant digits.

4. Zeros are significant when they serve as counters of the unit of measurement used. We have in 400.34 pounds the implied unit of measurement 0.01 of a pound. Here the total number of units is 40,034, and the significant digits are 4, 0, 0, 3, and 4.

There is one instance in which zeros are not significant. They are not significant when they serve as place holders only. For instance in 0.0070 inch, where the unit of measure is implied to be 0.0001 inch, there is a total of 70 units, and the significant digits are the last two on the extreme right, namely the 7 and the zero. The other three zeros, reading from left to right, locate the decimal point.

Zeros may or may not be significant when they stand to the right of significant digits but are on the left side of the decimal point. In 5640 gallons, if the unit of measurement is 10 gallons, the zero is not significant. However, if the unit of measurement is 1 gallon, the zero is significant. Numbers used in problems may have ambiguous zeros. From this point on, in this text, we shall state the unit of measure or the number of significant digits wherever ambiguity may occur.

4 | Rounding numbers

Frequently we have a tendency to retain more digits in numbers used in computations and in the results than are justified by the original data. The main reason for this practice of using unnecessary digits is that those working with the numbers do not know how many digits should be retained. They feel obliged to use all the digits given in the numbers, and they give the final result with much greater accuracy than the original measurements justify. This type of calculating is a waste of both time and energy.

For example, in finding the circumference of a circle with a radius 8.9 feet, we should not use 3.141592653 as the value for pi (π), even though we know that these are the first ten digits in pi. In discussing the multiplication of approximate numbers in a later portion of this chapter we shall see that the given radius of 8.9 feet justifies using only 3.14 for pi in the computing of the circumference of the circle in question.

We have already considered to some extent in an earlier chapter the "rounding of numbers." Let us now examine this topic further. Rounding a number is the process of dropping digits from the right end of a number. We replace with zeros the digits that are dropped only when it is necessary in order to keep the decimal point in its proper place. Rounding a number is done by one of the following rules:

1. If the digit to be dropped is less than 5, the digit on its immediate left that is to be kept should not be changed. Thus 85.642, which has five digits, becomes 85.64 when rounded to four digits.

2. If the digit to be dropped is larger than 5, the digit on its immediate left which is to be kept should be increased by 1. Now 85.647, which has five digits, when rounded to four digits becomes 85.65.

3. If the digit to be dropped is 5 and the digit to its immediate left is even, this digit on the immediate left, which is to be kept, is left unchanged. Thus 23.45, which has four digits, when rounded to three digits becomes 23.4.

4. If the digit to be dropped is 5 and the digit to its immediate left is odd, this digit on the immediate left, which is to be kept, should be increased by 1. Now 23.75, which has four digits, when rounded to three digits becomes 23.8.

5. If several digits are to be dropped from a number and the left-hand digit of those to be dropped is less than 5, the digit on its left (which is to be kept) should not be changed. If the left-hand digit of those to be dropped is 5 or larger, the digit on its immediate left (which is to be kept) should be increased by 1.

Thus 753.84739, which has eight digits, when rounded to four digits becomes 753.8. Also 234.68842, which has eight digits, when rounded to five digits becomes 234.69.

EXERCISE 1

In Problems 1 through 20, give the unit of measure that is implied

1. 729 pounds
2. 926 pounds
3. 3840 quarts
4. 7570 quarts
5. 4200 feet
6. 8500 feet
7. 56,792 feet
8. 359,462 feet
9. 47.6 miles
10. 667.8 miles
11. $83,729
12. 50,000 miles
13. $165,793
14. 180,000 miles
15. 682.07 inches
16. 8549.06 feet
17. 8.009 gallons
18. 47.0006 inches
19. $27\frac{3}{4}$ feet
20. $780\frac{3}{8}$ miles

In Problems 21 through 40, give the number of significant digits in each case.

21. 8756
22. 7846
23. 79,600
24. 86,509
25. 179,708
26. 32.009
27. 49,0068
28. 50.072
29. 0.6
30. 0.05
31. 0.007
32. 0.0063
33. 8,507,603
34. 93,708,007
35. 0.0608
36. 0.05003
37. 80,509.05
38. 60,000.007
39. 50.0805
40. 0.0000608

APPROXIMATE NUMBERS

B.

Round the following to two significant figures.

1. 87.6   2. 98.8   3. 76.4
4. 85.3   5. 7.59   6. 4.23

Round the following to three significant figures.

7. 534.7   8. 657.9   9. 458.3
10. 874.2   11. 984.6   12. 229.8

Round the following to four significant figures.

13. 6792.9   14. 7367.4   15. 486.73
16. 639.79   17. 73.046   18. 63.545

Round the following to five significant figures.

19. 2585.55 20. 40,569.3 21. 36,238.5
22. 90,005.5 23. 89.3478   24. 7.32555

Round each of the following to tenths.

25. 8.64   26. 9.75   27. 5.79
28. 4.32   29. 28.65   30. 17.76
31. 14.86   32. 19.85   33. 32.66
34. 129.12   35. 296.787   36. 5342.873
37. 7,969.86 38. 54,763.84 39. 46.77
40. 834.24

Round each of the following to hundredths.

41. 8.567   42. 4.638   43. 9.475
44. 7.543   45. 12.264   46. 25.645
47. 53.366   48. 74.394   49. 128.967
50. 396.859 51. 4532.543 52. 3647.687

5 | Addition and subtraction of approximate numbers

Results obtained by computing with approximate numbers are, of course, approximate. The accuracy of this resulting approximate number depends on the accuracy of each number appearing in the calculations. Let us set up rules for handling the results of these calculations. These rules should serve two purposes:

1. It is desirable to keep in the final result only those digits that are correct or likely to be correct. In other words, the results should

be written so as to be consistent with the practices already described in connection with approximate numbers.

2. It is well to eliminate all calculations that do not affect the final result. As we have pointed out earlier in this chapter, it is a complete waste of time in computing the area and circumference of a circle with radius 3.6 inches, to use an approximation for pi to ten decimal places.

Let us use the following practices in adding and subtracting approximate numbers:

1. Round each number so as to carry not more than one smaller place value (one more decimal place) than the term having the least precision.

2. Round the sum or difference of the terms in a set to the same number of places as the term having the least precision.

For example, add $437.563 + 98.4 + 36.5432 + 16.58 + 0.0567$. These become:

$$\begin{array}{r} 437.56 \\ 98.4 \\ 36.54 \\ 16.58 \\ 0.06 \\ \hline 589.14 \text{ or } 589.1 \end{array}$$

The term having the least precision is 98.4. It is precise to tenths. Thus, each of the other addends is rounded to hundredths, eliminating unnecessary computation. The sum of the rounded numbers is obtained according to the usual method of addition and then rounded to tenths in order to agree with the 98.4.

As another example, subtract 85.78 from 683.2187.

$$\begin{array}{r} 683.219 \\ 85.78 \\ \hline 597.439 \text{ or } 597.44 \end{array}$$

EXERCISE 2

Add the following approximate numbers.

1. $3.63 + 8.754 + 6.4 + 9.86$

APPROXIMATE NUMBERS

2. $15.78 + 9.4 + 16.87 + 4.649$

3. 28.65
    5.4
    0.738
    7.543
  469.73
    37.8

4. 7854.8
    29.653
      0.07
    698.287
     37.0
    946.395

5. Write in words each of the addends in Problem 3 and also the sum.

6. Write in words each of the addends in Problem 4 and also the sum.

7. 627.587
   −36.35

8. 879.3867
   −68.49

9. 7,563.5709
   −634.713

10. 86,579.36
    −542.7

11. 68,594.37
     −52.0

12. 869.873
   −15.09

13. Mr. Atherton travels the following distances each day during a week: 136.5 miles, 84.9 miles, 37.6 miles, 285.6 miles, and 17.6 miles. Find his total mileage for the week.

His average week's driving is 680 miles. By how much is his driving for this week below his average week?

14. The sides of a field are found by measurement to be 729.6 feet, 835.75 feet, 1036.4 feet, and 394.25 feet. Find the distance around the field.

15. The Blue Bonnet Dairy weighs in the following amounts of milk in a week. Find the total for the week: 3246.8 pounds, 3239.7 pounds, 3342.5 pounds, 3165.5 pounds, 3268.4 pounds, 3198.6 pounds, and 3310.8 pounds.

16. The following amounts of gasoline are purchased on a trip: 16.3 gallons, 10 gallons, 14.6 gallons, 12.8 gallons, 13.6 gallons, and 10.9 gallons. Find the total amount of gasoline bought.

17. The monthly rainfall in the southeastern United States measured as follows: 5.83 inches, 4.8 inches, 2.07 inches, 4.9 inches,

5.43 inches, 3.7 inches, 4.73 inches, 1.8 inches, 3.57 inches, 4.42 inches, 3.9 inches, and 2.84 inches. Find the total rainfall for the year. If the average yearly rainfall is 48.4 inches, how does the year in question compare with the average?

18. The hours of daylight in a week are measured for each day as follows: 12.7 hours, 12.34 hours, 12.6 hours, 12.79 hours, 12.53 hours, 12.42 hours, and 12.69 hours. Find the total number of daylight hours in the week.

19. The sides of a triangular field are measured as follows: 985.63 feet, 428.9 feet, and 726.543 feet. Find the distance around the field.

20. In Miss Brown's typewriting class 5 speed tests are given in a week. The lengths of these tests are as follows: 48.4 minutes, 39.6 minutes, 29.6 minutes, 27 minutes, and 17.8 minutes. What is the total time devoted to speed tests in a week?

6 | Multiplication and division of approximate numbers

Consider a rectangular strip of land 15.5 feet long and 25.5 feet wide. If we declare the area to be 395.25 square feet we are not being very realistic. The unit of measurement used for the sides does not justify the degree of refinement in measure that we have used to express the area. The long side of the rectangle could have varied from 25.45 feet to 25.55 feet. The short side could have varied from 15.45 feet to 15.55 feet. If the measurements had been 25.45 feet and 15.45 feet, which are the minimum possible lengths, the minimum area would be 393.20 square feet. If the measurements had been 15.55 feet and 25.55 feet, the maximum possible lengths, the maximum area would have been 397.30 square feet. We know that the true area of the rectangle falls somewhere between 393.20 square feet and 397.30 square feet. Let us notice that there are 4.1 square feet between the possible limits, and there is no justification in giving the computed area of 395.25 square feet any more accurately than the nearest square foot. This decision leads us to accept 395 square feet as the most probable area. It contains the same number of significant digits as the original linear measurements.

We shall follow two rules in the multiplication of approximate numbers.

APPROXIMATE NUMBERS   141

1. In finding the product of two approximate numbers, each of which has the same number of significant digits, keep the same number of significant digits in the product as are in the multiplier or the multiplicand.

2. In finding the product of two approximate numbers, one of which contains more significant digits than the other, round the more accurate of the numbers to be multiplied until it contains one more significant digit than the other before multiplying. The product should be rounded until it contains the same number of significant digits as appear in the less accurate of the numbers to be multiplied.

As we have already noted, an exact number differs from an approximate number in that an exact number represents a true count. We shall follow this rule in multiplying an exact number by an approximate number.

3. In finding the product of an exact number and an approximate number, keep in the product the same number of significant digits as appear in the given approximate number.

As an example of rule 1, multiply 16.57 by 43.12.

$$
\begin{array}{r}
16.57 \\
43.12 \\
\hline
3314 \\
1657 \\
4971 \\
6628 \\
\hline
714.4984
\end{array}
$$

This product rounds to 714.5.

As an example of rule 2, multiply 58.7921 by 16.3. Before multiplying, round the 58.7921 to 58.79.

$$
\begin{array}{r}
58.79 \\
16.3 \\
\hline
17637 \\
35274 \\
5879 \\
\hline
958.277
\end{array}
$$

This product will be rounded to 958.

To illustrate rule 3, multiply 358.62 by 15. In this example, we are considering that the 15 is an exact number and the 358.62 is an approximate number.

$$
\begin{array}{r}
358.62 \\
15 \\
\hline
179310 \\
35862 \\
\hline
5379.30
\end{array}
$$

This product rounds to 5379.3.

In solving a problem based on multiplying together *more than two approximate numbers*, find the product of two of the approximate numbers, round this partial product to one more digit than will be retained in the final answer, and use this rounded product with a third approximate number to find a new product. Continue the process until all the multiplications have been completed. Round the final product so that it contains the same number of significant digits appearing in the least accurate of the approximate numbers to be multiplied. For example, take $57.32 \times 14.6 \times 892.47$.

$$
\begin{array}{r}
57.32 \\
\times 14.6 \\
\hline
34392 \\
22928 \\
5732 \\
\hline
836.872
\end{array}
$$

We shall round this to 836.9. Now multiply 836.9 by 892.5, which is 892.47 rounded to four significant digits.

$$
\begin{array}{r}
892.5 \\
\times 836.9 \\
\hline
80325 \\
53550 \\
26775 \\
71400 \\
\hline
746{,}933.25
\end{array}
$$

This number will round to 747,000, the zeros being insignificant.

APPROXIMATE NUMBERS 143

Division of approximate numbers is carried on according to the following rules:

1. In finding the quotient of two approximate numbers, each of which has the same number of significant digits, divide the given dividend by the divisor until the quotient has one more significant digit than the divisor or dividend. The quotient is then rounded to the same number of significant digits as the divisor or dividend.

Consider 68.3 divided by 25.4.

$$\begin{array}{r} 2.688 \\ 25.4\overline{\smash{)}68.3.00} \\ 508 \\ \hline 1750 \\ 1524 \\ \hline 2260 \\ 2032 \\ \hline 2280 \\ 2032 \\ \hline \end{array}$$ This result rounds to 2.69.

2. In finding the quotient of two approximate numbers, one of which has more significant digits than the other, round the more accurate number until it has one more significant digit than the less accurate number. Divide the given dividend by the divisor until the quotient has one more significant digit than the less accurate number appearing in the dividend or the divisor. Then round the quotient so that it has the same number of significant digits as the less accurate number appearing in the dividend or in the divisor. Consider 65.238 divided by 4.7. Here the dividend, 65.238, is rounded to 65.2, in accordance with the foregoing rule.

Now:

$$\begin{array}{r} 13.8 \\ 4.7\overline{\smash{)}65.2.0} \\ 47 \\ \hline 182 \\ 141 \\ \hline 410 \\ 376 \\ \hline \end{array}$$

The 13.8 is rounded to 14, which is the final answer according to the rule just described.

3. In finding the quotient of an approximate and an exact number, divide the given dividend by the divisor until the quotient has one more significant digit than the given approximate number contains. Round the quotient so that it contains the same number of significant digits as the given approximate number contains.

As an illustration, consider the approximate number 5.873 divided by the exact number 15.

$$
\begin{array}{r}
0.39153 \\
15\overline{\smash{)}5.8730} \\
45 \\
\hline
137 \\
135 \\
\hline
23 \\
15 \\
\hline
80 \\
75 \\
\hline
50 \\
45 \\
\hline
\end{array}
$$

This quotient rounds to 0.3915.

EXERCISE 3

Multiply the following, assuming that all numbers are approximate unless otherwise indicated.

1. $2.8 \times 5.1 =$ 2. $3.7 \times 4.3 =$ 3. $59.4 \times 6.9 =$
4. $87.3 \times 4.6 =$ 5. $9.8 \times 36.7 =$ 6. $7.4 \times 42.9 =$
7. $18.6 \times 5.4 =$ 8. $21.6 \times 37.8 =$ 9. $68.4 \times 79.6 =$
10. $509.3 \times 3.4 =$ 11. $864.7 \times 5.6 =$ 12. $478.6 \times 47.2 =$
13. $829.7 \times 38.7 =$

14. $34 \times 79.83 =$ (Assume that the 34 is an exact number.)
15. $564.58 \times 15 =$ (Assume that the 15 is an exact number.)
16. $853 \times 54 =$ (Assume that both numbers are exact.)

APPROXIMATE NUMBERS —————————————————————145

17. $927 \times 46 =$ (Assume that both numbers are exact.)
18. $2346.57 \times 3.6 =$ **19.** $5693.78 \times 4.2 =$

20. Read the numbers that are to be multiplied together in Problem 18. Read the number that represents the product after this number has been rounded consistently according to significant digits.

21. Read the numbers that are to be multiplied together in Problem 19. Read the number that represents the product after this number has been rounded consistently according to significant digits.

22. $87.6 \times 23.5 \times 1.9 =$ **23.** $38.9 \times 6.2 \times 68.4 =$
24. $452.3 \times 78.4 \times 1.9 =$ *__25.__ $26.9 \times 33.5 \times 796.8 =$
*__26.__ $29.7 \times 3.84 \times .196 =$ *__27.__ $4.75 \times 26.7 \times .273 =$
*__28.__ $63.94 \times 49.8 \times 38.7 =$ *__29.__ $7.864 \times 37.6 \times 49.3 =$
30. $62.53 \times 2.6 \times 38.4 =$ *__31.__ $84.75 \times 26.8 \times 287.35 =$
*__32.__ $729.63 \times 83.7 \times 3.9 \times 27.85 \times 19.6 =$

Divide the following, assuming that all numbers are approximate unless otherwise indicated.

33. $57.8 \div 6.4 =$ **34.** $1.93 \div 7.8 =$
35. $17.84 \div 9.7 =$ **36.** $21.76 \div 8.5 =$
37. $17.8 \div 4.76 =$ **38.** $93.4 \div 14.5 =$
39. $753.4 \div 28.53 =$ **40.** $2573.7 \div 386.9 =$
41. $4.859 \div .75643 =$

42. $874.9 \div 63$ (Assume that 63 is an exact number.)

43. $2596.9 \div 124$ (Assume that 124 is an exact number.)

44. $795 \div 36.92 =$ (Assume that 795 is an exact number.)

45. $1243 \div 4.927 =$ (Assume that 1243 is an exact number.)

46. $49\overline{)764.35}$ (Assume that 49 is an exact number.)

47. $75\overline{)629.43}$ (Assume that 75 is an exact number.)

48. $847\overline{)2594.383}$ (Assume that 847 is an exact number.)

49. $926\overline{)4375.629}$ (Assume that 926 is an exact number.)

50. $8734\overline{)4976,854}$ (Assume that the dividend and divisor are both exact here.)

51. $24{,}976\overline{)7{,}856{,}982}$ (Assume that the dividend and divisor are both exact here.)

52. Read in words the divisor and dividend in Problem 50. Also read the quotient.

53. Read in words the divisor and dividend in Problem 51. Also read the quotient.

***54.** $17.549\overline{)359.78}$

***55.** $19.4684\overline{)307.6934}$

56. Read in words the divisor and dividend in Problem 55. Also read the quotient.

57. Read in words the divisor and dividend in Problem 54. Also read the quotient.

58. A plane on a certain flight used 76.4 gallons of gasoline per hour. If the flight lasted 3.2 hours, how many gallons of gasoline were used?

59. An acre contains 43,560 square feet. A piece of land is measured and the length is given as 574 feet and the width as 869 feet. How many acres does the plot contain, if we assume that the 574 and 869 are approximate numbers and the 43,560 is exact?

60. The lengths of the sides of a rectangle are measured with the use of a ruler graduated in tenths of an inch. The length of one side is read as 48.3 inches. The length of the other side is read as 56.9 inches. What is the smallest value that the area of this rectangle can have? What is the largest value that the area of the rectangle can have? Using the rules governing approximate numbers, give the area of this rectangle.

61. How many square feet are in the piece of land described in Problem 60?

In the following assume all numbers are approximate unless otherwise indicated.

62. Find the circumference of a circle of radius 7.9 inches, given $\pi = 3.1415926$.

63. A rectangular piece of carpet material is measured with a tape measure marked in units of $\frac{1}{8}$ foot. The carpet is said to be 16 feet wide and 27 feet long. What is the minimum area of carpet here?

64. What is the maximum area of carpet in Problem 63?

APPROXIMATE NUMBERS _____147

65. Ranches are frequently measured in sections. A section contains 640 acres. How many acres are in a ranch that is said to contain 104.6 sections? Assume that the 640 is an exact number.

66. Find the area of the circle described in Problem 62.

67. The diagonal of a square is equal to the length of the side multiplied by 1.4142136. What is the diagonal of a square whose sides are 19.7 inches?

68. The specific gravity of a substance is equal to the weight of a given volume of the substance divided by the weight of an equal volume of water. If we assume that water weighs 62.4 pounds per cubic foot and aviation gasoline weighs 45 pounds per cubic foot, what is the specific gravity of aviation gasoline?

69. How many pounds of milk containing 4.4% butterfat are needed to give 78 pounds of butterfat?

70. There is approximately 0.4535924277 kilogram in a pound. A boy weighs 176 pounds. How many kilograms does he weigh?

71. On a test Blake had 15 problems correct out of 20. What per cent of the problems did he have correct?

72. The live weight of a calf was 280 pounds. The dressed weight was 179 pounds. What was the per cent loss in weight?

73. The Willow Springs School playground has an area of 4683 square yards. If the width is 95 yards, find the length in yards.

74. A boat has a triangular sail with a base of 8 yards and a height of 10.5 yards. Find the area of the sail in square yards.

75. A turkey weighs 15.9 pounds. What will it cost at 59.4 cents a pound?

76. Multiply 759.6 feet by 0.64 and add 295.7 feet.

77. The Big Apple Warehouse sold 85 barrels of apples at $13.50 a barrel, and charged 21 cents a barrel for storage and a commission of $5\frac{3}{4}$%. Find the net proceeds of the sale.

78. How far is it around a piece of land represented by a rectangle 15 inches by 22 inches on a map to the scale of 1 inch = 0.8 mile?

Chapter 8

Elementary Algebra

1 | Introduction

Algebra is quite similar to arithmetic. It requires our careful concentration and persistent effort if we are to succeed with it. We have already observed that these traits are highly desirable as we have worked through the first seven chapters of this book. *Algebra may be said to be an extension of arithmetic in that, in algebra, letters are used to represent numbers. Thus algebra generalizes arithmetic.* For example, dog food sells for 20 cents per can. Two cans will cost 40 cents, 3 cans will cost 60 cents, and N cans will cost 20 times N cents or $20N$ cents, or $0.20N$ dollars. This gives us a formula that will work regardless of the number of cans of dog food purchased. Also, if N represents the number of cans of dog food that we buy in one week, and we buy four times as many cans the following week, then we buy 4 times N or $4N$ cans of dog food the following week.

Algebra utilizes the principles we have learned in arithmetic. Thus the four fundamental processes, indicated by the symbols $+$, $-$, $\times$, and $\div$ have the same meaning when applied to algebra that they had in arithmetic. Any combination of numbers and letters or of letters, and the four fundamental processes is called an algebraic expression. Parts of these expressions that are separated from each other by the signs $+$ or $-$ are known as terms. Thus $6a + 7b - 9m$ is an algebraic expression consisting of three terms. *When no sign is written in front of a term, the sign $+$ is understood.* The $6a$ in the above expression is understood to be plus.

If an algebraic expression contains one term it is called a simple expression. If it has two or more terms it is a multiple expression.

There are other common ways to describe algebraic expressions. *Those with one term are called monomials.* For example, $9a$ *is a monomial.* An expression with two terms, such as $7c + 8m$, *is called a binomial*, and an expression with three terms, such as $5x + 8y - 11z$, *is called a trinomial. The name multinomial refers to all multiple expressions, such as binomials and trinomials.*

When two or more quantities are multiplied together the result is known as the product of the quantities. When we use letters to represent numbers, there can be misunderstanding between the multiplication sign and the letters x and X. Consequently, we indicate multiplication in algebra by either (1) using a period halfway up between the letters, (2) writing the letters in succession with no symbol between them, or (3) writing each factor in parentheses. *Now the product of c and d may be written either as $c \cdot d$, or cd, or $(c)(d)$.*

Each of the quantities multiplied together to form a product is known as a factor of the product. Thus, 3, x, and y are the factors of the product $3xy$. Three is called the numerical factor, and x and y are literal factors.

The numerical factor in an algebraic expression is called the coefficient of the remaining factors. In the expression $5xy$, 5 is the coefficient, and xy is the literal coefficient of 5. The quantity x has the same meaning as $1x$. If the coefficient of a letter is not explicitly written, it is understood to be 1.

A power of a quantity is the result obtained when multiplying the quantity by itself any number of times. This multiplication is indicated by writing to the right and above the quantity a number that indicates the number of factors to be taken. Thus:

$b \times b$ is called the second power of b, and is written b^2
$b \times b \times b$ is the third power of b, and is written b^3
$b \times b \times b \times b$ is the fourth power of b, and is written b^4

The small number that indicates the power of any quantity is known as an exponent. Thus 2, 3, and 4 are the exponents of a^2, a^3, and a^4 respectively.

a^2 is generally read "a square"
a^3 is generally read "a cube"
a^4 is generally read "a to the fourth"

ELEMENTARY ALGEBRA

For example, $5a^4 + 4a^2 + 3a$ is read "five a to the fourth plus four a square plus three a."

The quantity x has the same meaning as x^1. If the power of a quantity is not explicitly written, it is understood to be 1.

For example, find the difference in meaning between $4x$ and x^4. By $4x$ we mean the result of multiplying 4 by x. By x^4, we mean the product of x and x and x and x. If $x = 2$, then $4x = 4 \cdot x = 4 \cdot 2 = 8$, but $x^4 = x \cdot x \cdot x \cdot x = 2 \cdot 2 \cdot 2 \cdot 2 = 16$.

Now, find the value of $5c^3$, when $c = 6$. But $5c^3 = 5 \cdot c \cdot c \cdot c = 5 \cdot 6 \cdot 6 \cdot 6 = 1080$.

As another example, find the value of $ab + bc - ac$ when $a = 3$, $b = 2$, and $c = 5$. Now $ab = 3 \cdot 2 = 6$; $bc = 2 \cdot 5 = 10$; and $ac = 3 \cdot 5 = 15$. Then $ab + bc - ac = 6 + 10 - 15 = 1$.

Also find $4(x + y)(x - y)$ where $x = 6$ and $y = 2$. The parentheses denote multiplication. Then

$$4(x + y)(x - y) = 4(6 + 2)(6 - 2) = 4(8)(4) = 128.$$

EXERCISE 1

Write each of the following in words. Then give the coefficient and the exponent of each letter.

1. $5x^2$
2. $4x^3$
3. $16x^2z^3$
4. $15a^3b^2$
5. $62a^2b^3c^4$
6. $13m^{10}$
7. $\frac{3}{5}x^3y^2$
8. $\frac{1}{3}p^2q^4$
9. $17a^3b^4c^5$
10. $54x^5y^9$

Write each of the following in exponent form.

11. $5xxxyy$
12. $14aabbb$
13. $xxxxxx$
14. $bbbbccc$
15. $64aabbb$
16. $\frac{5}{8}lllnn$
17. $196xxx^2x^4yy^2$
18. $46xxyyxzzz$
19. $108aaaa^2bbb^3$
20. $49xxx^5yyy^4y$

Find the value of the following when $a = 1$, $b = 2$, $c = 3$, and $d = 4$.

21. a^3b^3
22. a^2c^4
23. $5a^4 + b^2$
24. $18abc$

25. $8b + 3c - 4d$

26. $\dfrac{c + d}{14}$

27. $\dfrac{3b + 4d}{11}$

28. $\dfrac{d - b}{2}$

29. $ac + bd + 2cd$

30. $(4a + b - c)^4$

31. $\dfrac{c + d}{a} - \dfrac{2c - d}{b}$

32. $5b + 3(a + 4c - d)$

33. $(cd)^3 - (bd)^2$

34. $c^4 - \tfrac{1}{3}b^2 d$

35. $12d - 4(2a + 3b - c)$

36. $8(a + c)(d - a)$

37. $12(cd - ab)$

38. $\dfrac{48 a^3 b^2}{4 c^3 d^2}$

39. $\dfrac{2c + 3b - d}{2a + 3c - d}$

40. c^8

2 | Negative numbers

Negative numbers appear frequently in algebra. In order to illustrate the basic relation between positive and negative numbers, let us consider the positive numbers as so many points along a line, each a measured distance to the right of a reference point called

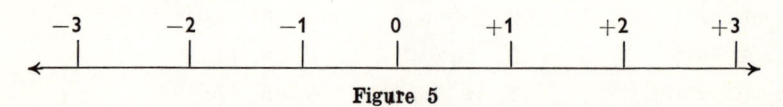

Figure 5

zero (0). We shall extend the line to the left of the zero point. We can measure off on the line to the left the same distances from zero that we have measured to the right. Numbers to identify these points on the left are known as *negative numbers*. These numbers are written with a minus sign (—) in front of them, in contrast with positive numbers, which are preceded by a plus sign (+). A number written without a sign is considered a positive number.

When we are thinking only in terms of measured distances, regardless of direction, we use *absolute values*. The absolute value of a number disregards the sign attached to the number. We designate absolute value of a number m by enclosing it between two parallel bars $|m|$. Thus $|-5| = |+5| = 5$.

ELEMENTARY ALGEBRA ────────────────────────────── **153**

The following are *rules* for *adding and subtracting signed numbers*.

1. When the addends have like signs:
   (*a*) Add the absolute values of the addends.
   (*b*) Place the common sign before the sum in step *a*.
2. When the addends have unlike signs:
   (*a*) Subtract the smaller of the two absolute values from the larger.
   (*b*) Place the sign of the addend with the larger absolute value in front of the difference of step *a*.

For example:
1. Where the signs are alike:
   (*a*) $(+3) + (+9) = + |3 + 9| = +12$ or 12
   (*b*) $(-4) + (-7) = - |4 + 7| = -11$
2. Where the signs are unlike:
   (*a*) $(-7) + (+4) = - |7 - 4| = -3$
   (*b*) $(+16) + (-10) = + |16 - 10| = +6$ or 6

To subtract two signed numbers:

(*a*) Change the sign of the subtrahend (this is the number to be subtracted).

(*b*) Add the changed subtrahend and the minuend by following the rule previously given for the addition of two signed numbers.

For example:
1. Subtract: $(+9) - (+3)$
   (*a*) $(+3)$ changes to (-3)
   (*b*) $(+9) + (-3) = +6$ or 6
2. Subtract: $(+7) - (-3)$
   (*a*) (-3) changes to $(+3)$
   (*b*) $(+7) + (+3) = +10$ or 10
3. Subtract: $(-15) - (+8)$
   (*a*) $(+8)$ changes to (-8)
   (*b*) $(-15) + (-8) = -23$

The following *rule* applies to *multiplying signed numbers*.

1. When the signs of two numbers are alike:
   (*a*) Obtain the product of the absolute values of the numbers.
   (*b*) Place a plus sign before the product in step *a*.
2. When the signs of the two numbers are not alike:
   (*a*) Obtain the product of the absolute values of the numbers.
   (*b*) Place a minus sign before the product in step *a*.

For example:
1. Where the signs are alike:
   (a) $(+6) \times (+4) = + |6 \times 4| = +24$ or 24
   (b) $(-8) \times (-5) = + |8 \times 5| = +40$ or 40
2. Where the signs are unlike:
   (a) $(+7) \times (-9) = - |7 \times 9| = -63$
   (b) $(-8) \times (+4) = - |8 \times 4| = -32$

The following *rule* applies to *dividing signed numbers*.
1. When the signs of the divisor and dividend are alike:
   (a) Divide the absolute value of the divisor into the absolute value of the dividend.
   (b) Place a plus sign before the quotient obtained in the step above.
2. When the signs of the divisor and dividend are not alike:
   (a) Divide the absolute value of the divisor into the absolute value of the dividend.
   (b) Place a minus sign before the quotient obtained in the step above.

For example:
1. Where the signs are alike:
   (a) $(+8) \div (+2) = + |8 \div 2| = +4$ or 4
   (b) $(-18) \div (-6) = + |18 \div 6| = +3$ or 3
2. Where the signs are unlike:
   (a) $(-18) \div (+6) = - |18 \div 6| = -3$
   (b) $(+24) \div (-3) = - |24 \div 3| = -8$

EXERCISE 2

Add the following:

1. $(-12) + (-9)$
2. $(-7) + (-15)$
3. $(-19) + (-14)$
4. $(-18) + (-9)$
5. $(+19) + (-4)$
6. $(+23) + (-15)$
7. $(+27) + (-5)$
8. $(+34) + (-14)$
9. $(-16) + (+12)$
10. $(-13) + (+8)$
11. $(-19) + (+13)$
12. $(-33) + (+15)$
13. $(+143\tfrac{3}{8}) + (-42\tfrac{1}{4})$
14. $(+367\tfrac{3}{4}) + (-238\tfrac{5}{16})$
15. $(-526\tfrac{7}{8}) + (+238\tfrac{3}{5})$
16. $(-379\tfrac{5}{8}) + (+136\tfrac{3}{4})$

ELEMENTARY ALGEBRA —————————————————155

Subtract the following:
17. $(+19) - (-4)$
18. $(+37) - (-15)$
19. $(+46) - (+23)$
20. $(+98) - (+64)$
21. $(-79) - (-14)$
22. $(-84) - (-63)$
23. $(-96) - (-53)$
24. $(-137) - (-48)$
25. $(-48\frac{5}{8}) - (-110\frac{3}{4})$
26. $(-38.7) - (+15.07)$
27. $(-148.7) - (+307.9)$
28. $(+68.5) - (-287.4)$

Multiply the following:
29. $(+8)(+4)$
30. $(+12)(+7)$
31. $(-5)(-8)$
32. $(-15)(-4)$
33. $(-17)(-6)$
34. $(-19)(-8)$
35. $(+35)(-5)$
36. $(-14)(+19)$
37. $(-42)(+8)$
38. $(+64)(-15)$
39. $(-3.54)(+17.9)$
40. $(-68\frac{1}{4})(+14\frac{3}{4})$
41. $(-168\frac{3}{8})(-142\frac{3}{4})$
42. $(+265\frac{3}{4})(-19\frac{1}{2})$

Divide the following:
43. $(+16) \div (+4)$
44. $(-32) \div (-8)$
45. $(+81) \div (-9)$
46. $(-34) \div (+17)$
47. $(-126) \div (-21)$
48. $(+18.2) \div (-9.1)$
49. $(-6.48) \div (+32.4)$
50. $(-12\frac{1}{4}) \div (-6\frac{1}{2})$
51. $(+0.0078) \div (+0.39)$
52. $(-15.3) \div (+5.1)$

3 | Solving simple linear equations

In our work in arithmetic we have solved many problems dealing with "unknown numbers." Using the four fundamental operations, addition, subtraction, multiplication, and division, these problems appeared as one of four basic types, as described below.

Type 1. What number increased by 5 equals 15? This could be written $? + 5 = 15$.

In algebra, we let x (or any other letter) represent the unknown quantity that we seek. An equation is a statement of equality between two quantities. Now we may write the equation referred to just above, replacing the question mark by x.

Then $x + 5 = 15$.

A statement that two expressions are equal such as we see in

Type 1 above, is called an equation. The terms on the left of the sign of equality are called the left member or left side of the equation. The terms on the right are called the right member or right side of the equation. A number that reduces both members to the same number when it is substituted for the unknown in an equation is said to be a root or solution of the equation. It also satisfies the equation.

The solution of the equation

$$x + 5 = 15$$

is

$$x = 10$$

Thus 10 is a root of the equation $x + 5 = 15$. To see if 10 satisfies the equation, we replace x by 10.

Then

$$10 + 5 = 15$$

or

$$15 = 15$$

Type 2. What number decreased by 5 equals 15? This may be written

$$? - 5 = 15$$

or

$$x - 5 = 15$$

The solution of this equation is $x = 20$.

Type 3. What number multiplied by 5 equals 15? This may be written

$$(?)(5) = 15$$

or

$$5x = 15$$

The solution of this equation is $x = 3$.

Type 4. What number divided by 5 equals 15? This may be written

$$\frac{?}{5} = 15$$

or

$$\frac{x}{5} = 15$$

The solution of this equation is $x = 75$.

ELEMENTARY ALGEBRA _____**157**

    The following rule is basic in the handling of equations: *If equal numbers are increased, decreased, multiplied, or divided by the same number, the resulting numbers will be equal. We shall exclude division by zero.*

    Consider the equation $15 = 15$. It makes no difference what number we either add to or subtract from both members of this equation. The resulting numbers will always be equal.

Given: $\quad\quad\quad 15 = 15$
$\quad\quad\quad\quad 15 + 2 = 15 + 2$, adding 2 to both sides
$\quad\quad\quad\quad 17 = 17$, the results are equal

Given: $\quad\quad\quad 15 = 15$
$\quad\quad\quad\quad 15 - 2 = 15 - 2$, subtracting 2 from both sides
$\quad\quad\quad\quad 13 = 13$, the results are equal

Given: $\quad\quad\quad 15 = 15$
$\quad\quad\quad\quad 15 - 18 = 15 - 18$, subtracting 18 from both sides
$\quad\quad\quad\quad -3 = -3$, the results are equal

Given: $\quad\quad\quad 15 = 15$
$\quad\quad\quad\quad 15 \cdot 5 = 15 \cdot 5$, multiplying both sides by 5
$\quad\quad\quad\quad 75 = 75$, the results are equal

Given: $\quad\quad\quad 15 = 15$
$\quad\quad\quad\quad \dfrac{15}{5} = \dfrac{15}{5}$, dividing both members by 5
$\quad\quad\quad\quad 3 = 3$, the results are equal

Given: $\quad\quad\quad x + 5 = 15$
$\quad\quad\quad\quad x + 5 - 5 = 15 - 5$, subtracting 5 from both sides
$\quad\quad\quad\quad x = 10$, simplifying

Given: $\quad\quad\quad x - 5 = 15$
$\quad\quad\quad\quad x - 5 + 5 = 15 + 5$, adding 5 to both sides
$\quad\quad\quad\quad x = 20$, simplifying

Given: $\quad\quad\quad 5x = 15$
$\quad\quad\quad\quad \dfrac{5x}{5} = \dfrac{15}{5}$, dividing both sides by 5
$\quad\quad\quad\quad x = 3$, simplifying

Given: $\quad\quad\quad \dfrac{x}{5} = 15$
$\quad\quad\quad\quad \dfrac{x}{5} \cdot 5 = 15 \cdot 5$, multiplying both sides by 5
$\quad\quad\quad\quad x = 75$, simplifying

    Once we arrive at what apparently is the solution to an equation, it is important to see if this solution satisfies the equation. This procedure is called checking the solution. A value of the unknown, which satisfies an equation, is called a *solution* of the equation.

Solve and check the equation $x + 7 = 12$.

$x = 12 - 7$, subtracting 7 from both sides

$x = 5$, simplifying

To check, we replace the x in the original equation by 5. Then:

$$5 + 7 = 12$$

and

$$12 = 12, \text{ simplifying}$$

Solve and check the equation $8x = 48$.

$x = \dfrac{48}{8}$, dividing both sides by 8

$x = 6$, simplifying

To check, we replace the x in the original equation by 6. Then:

$$8 \cdot 6 = 48$$

$$48 = 48, \text{ simplifying}$$

Solve and check the equation $x - 4.7 = 5.3$ (decimals).

$x = 5.3 + 4.7$, adding 4.7 to both sides

$x = 10.0$, simplifying

To check, we replace the x in the original equation by 10.0. Then:

$$10.0 - 4.7 = 5.3$$

$$5.3 = 5.3, \text{ simplifying}$$

Solve and check the equation $\frac{x}{7} = 4$.

$x = 7 \cdot 4$, multiplying both sides by 7

$x = 28$, simplifying

To check, we replace the x in the original equation by 28.

Then:

$$\frac{28}{7} = 4$$

$4 = 4$, simplifying

Solve and check $5x + 4 = -21$.

$5x = -25$, adding -4 to each side

$x = -5$, simplifying

In checking $x = -5$ as the solution of $5x = -25$, we have $(5)(-5) = -25$.

$-25 = -25$, simplifying

Solving word problems is fascinating. If we take each part of the problem, study it carefully, and then put together our various conclusions, we can be quite successful with word problems in algebra.

For example, one number is 7 more than three times the other, and their sum is 35. Find the numbers.

We shall begin by letting x equal one number. The other number will be $7 + 3x$. Adding the expressions for the two numbers we have $x + 7 + 3x$, which equals 35. Then solving $x + 7 + 3x = 35$, we have $4x = 35 - 7$ or $4x = 28$. Then $x = 7$ is one number. The other number is $7 + 3x$ or $7 + 3(7)$ or $7 + 21$ or 28.

To check we go back to the original wording of the problem. We have found that one number is 28. This is 7 more than three times 7 which is the other number. Also the sum of the numbers 7 and 28 is 35.

Another example of a word problem is: A man is six times as old as his daughter and his daughter is five years younger than his son. If their combined ages total 77 years, how old is each?

We shall begin by letting

$x = $ the age of the son

Then

$x - 5 = $ the age of the daughter

$6(x - 5) = $ the age of the father

Since the combined ages total 77,

    age of   age of   age of
     son   daughter   father

$$x + x - 5 + 6(x - 5) = 77$$
$$x + x - 5 + 6x - 30 = 77$$
$$8x = 112$$
$$x = 14 \text{ son's age}$$
$$x - 5 = 9 \text{ daughter's age}$$
$$6(x - 5) = 54 \text{ father's age}$$

We see that according to the ages we have found, the daughter is five years younger than the son, and the father is six times as old as the daughter. Also, the sum of the ages 14, 9, and 54 totals 77.

EXERCISE 3

Solve and check the following equations.

1. $x + 4 = 7$
2. $x - 5 = 9$
3. $x + 9 = 15$
4. $x + 8 = 3$
5. $x + 16 = 11$
6. $x + 9 = 5$
7. $x + 13 = 22$
8. $x + 14 = 19$
9. $6x = 24$
10. $8x = 48$
11. $9x = 45$
12. $11x = 66$
13. $x - 5 = 7\frac{1}{2}$
14. $x + 12 = 14\frac{3}{8}$
15. $x - 19 = 8\frac{1}{4}$
16. $x - 15 = 17\frac{3}{5}$
17. $\dfrac{x}{8} = 5$
18. $\dfrac{x}{5} = 6$
19. $\dfrac{x}{5} = 9\frac{1}{7}$
20. $\dfrac{x}{8} = 6\frac{1}{5}$
21. $x - 5\frac{1}{4} = 3\frac{1}{2}$
22. $x + \dfrac{3}{8} = 4\frac{1}{4}$
23. $x + 2\frac{3}{5} = 7\frac{3}{4}$
24. $x - 7\frac{3}{4} = 18\frac{1}{2}$
25. $\dfrac{x}{9} = 5\frac{1}{4}$
26. $\dfrac{x}{6} = \dfrac{5}{6}$

27. $\dfrac{x}{8} = 3\tfrac{2}{3}$ **28.** $\dfrac{x}{9} = 5\tfrac{2}{5}$

29. $4x = 16.4$ **30.** $7x = 70.7$

31. $x + 5.64 = 8.72$ **32.** $x + 8.35 = 12.46$

33. $7x - 14 = 5x$ **34.** $4x + 9 = x$

35. $\dfrac{5x}{8} + 9 = \dfrac{x}{4} + 8$ **36.** $\dfrac{7x}{5} - 8 = \dfrac{x}{7} + 9$

37. $4\tfrac{1}{5}x - 9 = 3x + 7$ **38.** $13x - 10 = 9x + 4$

39. $\dfrac{5x}{4} + 12 = 19$ **40.** $\dfrac{x}{3} + \dfrac{x}{4} = 12$

41. $\dfrac{4x}{5} + \dfrac{x}{4} = 31$ **42.** $3\tfrac{3}{4}x + 8 = 1\tfrac{3}{8}x - 4$

43. $12 - 4(3x - 2) = 16(3x + 7) - 48$

44. $16 + 5(3x - 7) = 14(2x - 8) + 25$

45. $63 + 3(4x + 5) = 19(7x - 13) + 16$

46. $5(4x - 9) - 6(2x - 1) = 13(4x - 7) + 3(8 - 3x)$

47. 60% of what number equals 48?

48. One number is 12 more than another and their sum is 168. Find the numbers.

49. Five times what number is 91 more than two-thirds of that number?

50. The sum of four consecutive odd integers is 184. Find the integers. (Hint: Let the first odd number be n; the next will then be $n + 2$, and the third will be $n + 4$.)

51. Eight less than six times a certain number is the same as four more than three times the number. Find the number. (Hint: Let $x =$ the number; then eight less than six times the number is $6x - 8$.)

52. The larger of two numbers exceeds three times the smaller by 2. If the sum of the numbers is 86, what are the numbers?

53. A team played a total of 166 games and won 42 games more than it lost. How many games did it win?

54. A man is four times as old as his daughter and his daughter is five years older than his son. If their combined ages total 43 years, how old is each?

55. Mr. Branch and Mr. Patton divided $40,000, Mr. Patton receiving four times as much as Mr. Branch. How much did each receive?

56. A piece of wire 10 feet long is cut into 3 pieces, so that one piece is four times as long as the shortest piece and the longest piece is five times as long as the shortest piece. Find the length of the 3 pieces.

57. Three-fifths of a number is equal to six more than one-tenth of the number. Find the number.

58. The sum of three consecutive integers is 102. Find the integers.

59. Seventy-eight dollars are to be divided among 6 boys and 8 girls, so that each of the boys receives three times as much as each of the girls. How much does each receive?

60. Jerome is 8 years older than his brother, and the sum of their ages is 96 years. How old is the brother?

61. Janie spent three-fourths of her money for a purse and one-sixth of her money for a pen. The remaining $2.15 she saved. How much money did she have in the first place?

4 | Formulas

A *formula* is an *equation* in which the letters refer to definite quantities. To find the value of any letter (also known as variable) in a formula when the values of the other letters are given, write down the formula, substitute the values that are given for the variables, and carry out the necessary operations.

Example 1

In the formula $p = 2l + 2w$, find the value of p when $l = 5$ and $w = 3$.

Solution:

$$p = 2l + 2w$$
$$p = 2 \cdot 5 + 2 \cdot 3$$
$$p = 10 + 6$$
$$p = 16$$

Answer: $p = 16$

Example 2

Find the value of l when $A = 63$ and $w = 9$, using the formula $A = lw$.

Solution:
$$A = lw$$
$$63 = l \cdot 9$$
$$\frac{63}{9} = \frac{l \cdot 9}{9}$$
$$7 = l$$
$$l = 7$$

Answer: $l = 7$

EXERCISE 4

Find the value of:

1. $2(a - b)$ when $a = 9$ and $b = 6$.
2. $\frac{a}{b}$ when $a = 18$ and $b = 3$.
3. $5x - 2y$ when $x = 2$ and $y = 3$.
4. bh when $b = 8$ and $h = 6$.
5. πd when $\pi = 3.14$ and $d = 4$.
6. lwh when $l = 3$, $w = 4$, and $h = 5$.
7. $2\pi r$ when $\pi = \frac{22}{7}$ and $r = 14$.
8. $2\pi rh$ when $\pi = 3.14$, $r = 6$, and $h = 8$.
9. $b + 3c$ when $b = 4$ and $c = 8$.
10. $2l + 2w$ when $l = 42$ and $w = 64$.
11. $p - i$ when $p = 648$ and $i = 237$.
12. prt when $p = 800$, $r = 0.06$, and $t = 2$.
13. prt when $p = 1000$, $r = .05$, and $t = \frac{1}{2}$.
14. $\frac{1}{2}cd$ when $c = 16$ and $d = 8$.
15. $\frac{1}{4}\pi d^2$ when $p = 3.14$ and $d = 10$.
16. $\frac{4}{3}\pi r^3$ when $\pi = \frac{22}{7}$ and $r = 7$.
17. $\frac{1}{3}\pi r^2 h$ when $\pi = 3.14$, $r = 6$, and $h = 12$.
18. $h^2 - a^2$ when $h = 12$ and $a = 8$.

19. $p + prt$ when $p = 900$, $r = 0.06$, and $t = \frac{1}{3}$.
20. $2\pi rh + 2\pi r^2$ when $\pi = 3.14$, $r = 12$, and $h = 18$.
21. A when $b = 9$ and $h = 16$. Formula: $A = bh$.
22. c when $\pi = \frac{22}{7}$ and $d = 70$. Formula: $c = \pi d$.
23. v when $l = 23$, $w = 14$, and $h = 9$. Formula: $v = lwh$.
24. A when $s = 20$. Formula: $A = s^2$.
25. p when $l = 128$ and $w = 63$. Formula: $p = 2l + 2w$.
26. l when $a = 8$, $n = 14$, and $d = 6$. Formula: $l = a + (n-1)d$.
27. s when $p = 16$. Formula: $p = 4s$.
28. r when $d = 54$. Formula: $d = 2r$.
29. l when $A = 48$ and $w = 8$. Formula: $A = lw$.
30. d when $c = 25.12$ and $\pi = 3.14$. Formula: $c = \pi d$.
31. h when $v = 1271.7$, $\pi = 3.14$, and $r = 9$. Formula: $v = \pi r^2 h$.
32. $c^2 + d^2$ when $c = 1.1$ and $d = 2.3$.
33. $v + gt$ when $v = 180$, $g = 32$, and $t = 9$.
34. w when $v = 2646$, $l = 21$, and $h = 14$. Formula: $v = lwh$.
35. r when $A = 3300$, $\pi = \frac{22}{7}$, and $h = 35$. Formula: $A = 2\pi rh$.
36. h when $A = 5720$, $\pi = \frac{22}{7}$, and $d = 28$. Formula: $A = \pi dh + \frac{1}{2}\pi d^2$.
37. h when $v = 243$ and $B = 81$. Formula: $v = \frac{Bh}{3}$.
38. r when $A = 13{,}000$, $p = 1000$, and $t = 8$. Formula: $A = p + prt$.

5 | Some formulas from commercial algebra

Let us now consider some of the more basic formulas that have frequent application in the business world.

When we rent a house for the family we expect to pay for the use of the house. When we borrow money we expect to pay for the use of the money. The money we pay for the use of the house is called *rent*. The amount paid for the use of money is called *interest*.

Suppose you borrow $800 for 3 months at 8%. If you borrowed $800 for one year at 8% you would pay

$$\$800 \times 0.08 = \$64.00$$

But 3 months are one-fourth of a year. Now, $\frac{1}{4} \times \$64.00 = \$16.00$.

ELEMENTARY ALGEBRA ——————————————————————————**165**

This is the amount of interest you owe on the $800. A rather simple way to look at the problem is to take the formula $I = prt$. Here I represents the interest, p the principal, r the rate, and t *the time in years.*

Thus, in the example to which we have just referred, $p = \$800$, $r = 8\%$ or 0.08 and $t = \frac{1}{4}$. Now in $I = prt$,

$$I = \left(\cancel{800}^{\,8}\right)\left(\cfrac{8}{\cancel{100}}^{\,2}\right)\left(\cfrac{1}{\cancel{4}}_{\,1}\right)$$

or $I = \$16.00$.

In repaying the loan of $800 plus interest you will pay a total of $816. It is convenient to express this through the formula $A = P + I$, in which A represents the final amount to be paid, P is the principal, and I the interest. In the illustration of borrowing $800 for 3 months at 8%, we have $P = \$800$, $I = \$16$ and A, the total amount to be repaid, is $A = \$800 + \$16 = \$816$.

The formula $I = prt$ becomes $t = \frac{I}{pr}$ upon solving for t.

If we wish to determine how long $1000 is borrowed at 6% when $20 interest is due at the time the loan is repaid, then $p = \$1000$, $r = 0.06$, and $I = \$20$.

$$t = \frac{20}{(1000)(.06)} = \frac{20}{60} = \frac{1}{3}$$

Since t represents the time in years, we have one-third of a year or 4 months as the length of time for which the $1000 is borrowed.

The formula $I = prt$ becomes $r = \frac{I}{pt}$ when solved for r. Suppose we have agreed to pay $40 interest on $2000, which we have borrowed for 6 months. Find the interest rate we are paying. Here $I = \$40$, $p = \$2000$, and $t = 6$ months.

Now

$$r = \frac{40}{(2000)(\frac{1}{2})} = \frac{\cancel{40}}{\cancel{1000}} = 0.04 \text{ or } 4\%$$

which is the rate of interest we are paying.

The formula $D = snd$ is another important one. Let us examine a situation in which it applies. Suppose you want to borrow $600 for 3 months from a bank that charges 8%. Banks and other lending agencies usually *charge interest in advance.* This means that the bank will not hand you a check for $600, which is the amount

you asked to borrow. Instead they take 8% of $600 for 3 months.
$$\frac{8}{\underset{\scriptscriptstyle 100}{}} \times \underset{\scriptscriptstyle 600}{\overset{6}{}} = \$48,$$ the interest for one year. Three months equals $\frac{1}{4}$ of a year.

$$\frac{1}{\underset{\scriptscriptstyle 4}{}} \times \$48 \overset{12}{=} \$12$$

This is the interest on $600 for 3 months at 8%. The bank gives you $588. Three months later you will be expected to pay the bank $600. Thus the bank is actually charging *interest in advance*. *Bank discount* is a business term for *interest in advance*.

Let us apply the formula $D = snd$ to the problem of borrowing $600 for 3 months from a bank that charges 8%. In this formula s represents the sum to be borrowed, in this case $600; n represents the time (*in years*), and d represents the discount rate (interest rate). Thus $n = \frac{1}{4}$ and $d = 0.08$. D equals the bank discount.

Now:

$$D = \left(\underset{\scriptscriptstyle 600}{\overset{6}{}}\right)\left(\frac{1}{\underset{\scriptscriptstyle 4}{}}\right)\left(\frac{\overset{2}{\cancel{8}}}{\underset{\scriptscriptstyle 100}{}}\right) = \$12$$

which is the bank discount.

As pointed out previously the bank pays out $588 in the loan transaction but expects to be paid back $600. In charging interest in advance, the bank actually collects a higher rate of interest than appears in the terms of loan.

We shall now take the same situation as before. Assume that we want to borrow $600 for 3 months from a bank that charges 8%. The sum of money we actually receive from the bank is $588. The amount of interest we pay is $12. Let us find the interest rate which is being charged.

Using $r = \frac{I}{pt}$, we have

$$r = \frac{12}{(588)(\frac{1}{4})} = \frac{12}{\frac{588}{4}} = \frac{48}{588} = 0.0816 \text{ or } 8.2\%$$

whereas the charge stated in the loan was 8%.

The policy of bank discount should be thoroughly understood by all who borrow money.

Ordinary and Exact Interest. In computing interest when the time is given in months, we take each month as $\frac{1}{12}$ of a year. If the time

ELEMENTARY ALGEBRA _____**167**

is given in days, we may take each day to be $\frac{1}{365}$ of a year as there are 365 days in a year. It is frequently more convenient, however, and in most business transactions it is the custom, to consider a day as $\frac{1}{360}$ of a year. Actually, since the denominator 365 is larger than the denominator 360, a smaller interest payment is made to the lender when 365 is used. Naturally the 360-day base is preferred by those who lend money.

Interest computed on the basis of a 360-day year is called *ordinary interest*. Interest computed on the basis of 365 days to the year is called *exact interest*. We shall use *ordinary interest* in all computations in this text involving simple interest.

Frequently in working with financial problems we find the time is not given as a number of days or months but we are asked to find the interest on a given principal from one date to another, say from June 12, 1971 to October 15, 1971. There are two methods we can use to find the time. One is the method of *approximate time*. This method works on the basis of 30 days to the month or 360 days to the year. Let us find the approximate time between June 12, 1971 and October 15, 1971.

	Year	Month	Day
October 15, 1971 =	1971	10	15
June 12, 1971 =	1971	6	12
Subtracting, we find:		4	3

Four months and 3 days or 123 days is the approximate time between June 12, 1971 and October 15, 1971.

The method of *exact time* is based on the exact number of days in each month. Thus the total number of days in a year is 365, except in leap years when there are 366.

Let us find the exact time between June 12, 1971 and October 15, 1971. There are 18 days left in June, 31 in July, 31 in August, 30 in September, and 15 in October. This total gives us 125 days. Notice that we counted the last day but not the first one. Also observe that by the method of *approximate time* we found 123 days between June 12, 1971 and October 15, 1971, whereas by the method of *exact time* we found 125 days. A person lending money is consequently smart to use *ordinary interest and exact time*.

For example, Jack Jones borrows $5000 from Harry Schmidt on April 16, 1972 and agrees to pay it back with 8% interest on November 19, 1972. Find how much Jones owes Schmidt on

November 19, 1972. We shall compute this sum on the basis of *ordinary interest* and *exact time*. We use $I = prt$ or

$$I = p \times r \times \frac{\text{exact time between dates}}{360}$$

We shall now find the exact time between April 16, 1972 and November 19, 1972. From April 16 to April 30 is 14 days. May has 31 days; June 30; July 31; August 31; September 30; October 31; and November 19. The total is 217 days.

Now:

$$I = \cancel{5000}^{50} \times \frac{\cancel{8}^{2}}{\cancel{100}} \times \frac{217}{\cancel{360}_{9}}$$

$I = \$241.11$, the amount of interest Jones owes Schmidt.

Jones also owes the $5000 principal he has borrowed. A convenient relationship is $A = P + I$, or amount equals principal plus interest. Now Jones owes Schmidt $5000 + $241.11, or $5241.11.

EXERCISE 5

In the following time periods find the approximate time between dates.

1. February 15, 1968 and May 24, 1968
2. March 12, 1969 and July 24, 1969
3. June 14, 1970 and October 8, 1970
4. August 18, 1971 and December 5, 1971
5. January 16, 1972 and November 10, 1972
6. April 28, 1970 and July 16, 1970
7. August 16, 1972 and November 21, 1972
8. May 11, 1973 and October 27, 1973
9. July 27, 1970 and October 16, 1970
10. August 15, 1971 and February 10, 1972
11. September 26, 1970 and May 24, 1971
12. July 19, 1968 and March 14, 1969
13. February 14, 1971 and June 10, 1971

14. June 23, 1972 and November 18, 1972
15. October 17, 1971 and April 10, 1972
16. November 19, 1972 and May 19, 1973

In the following find the exact time between dates.

17. September 17, 1972 and March 23, 1973
18. August 15, 1971 and December 24, 1971
19. June 25, 1972 and February 16, 1973
20. September 14, 1970 and April 10, 1971
21. May 10, 1971 and November 5, 1971
22. August 14, 1968 and April 20, 1969
23. November 29, 1969 and July 14, 1970
24. December 28, 1970 and October 11, 1971
25. October 15, 1972 and April 14, 1973
26. September 18, 1971 and January 14, 1972
27. July 28, 1972 and March 31, 1973
28. October 16, 1970 and May 15, 1971
29. June 16, 1971 and December 28, 1971
30. September 17, 1968 and May 24, 1969
31. December 30, 1969 and August 19, 1970
32. January 14, 1970 and November 16, 1971

In Problems 33 through 42 find the ordinary interest for the exact time.

33. Principal $800, rate 7%, January 22, 1968 to November 25, 1968
34. Principal $1500, rate 8%, July 9, 1969 to December 23, 1969
35. Principal $340, rate $7\frac{1}{2}$%, August 28, 1970 to February 24, 1971
36. Principal $3680, rate $9\frac{1}{4}$%, September 10, 1970 to August 18, 1971
37. Principal $2160, rate $8\frac{3}{4}$%, July 17, 1970 to May 26, 1971
38. Principal $3640, rate $7\frac{1}{2}$%, 7 months
39. Principal $48,000, rate $4\frac{1}{2}$%, 90 days
40. Principal $21,000, rate 4%, 30 days
41. How long will it take the interest on $3400 at 5% to amount to $78.40?

42. How long will it take the interest on $4600 at $4\tfrac{1}{2}\%$ to amount to $164.60?

43. At what rate will the interest on $1900 for 100 days be $36.85?

44. At what rate will the interest on $4400 for 140 days be $119.78?

45. What principal will amount to $2168.75 in 95 days at $5\tfrac{1}{2}\%$?

46. The interest on a certain sum for 130 days at $5\tfrac{1}{2}\%$ was $58.40. Find the sum.

47. Ray Johnson borrows $1800 from a bank whose discount rate is $7\tfrac{1}{2}\%$. If he gives the bank a 120-day note, how much money does he receive from the bank? How much does he repay the bank at the end of 120 days?

48. What principal will amount to $5400 in 100 days at $4\tfrac{1}{2}\%$?

49. Shannon Brand went to the Third National Bank and signed a 120-day note with a face value of $4500. Interest in advance at $5\tfrac{1}{2}\%$ was deducted. How much did Brand receive on the loan? How much does he pay the Third National Bank at the end of 120 days?

50. Jane Ray borrows $2100 from a bank whose discount rate is $8\tfrac{1}{2}\%$. If she gives the bank a 150-day note, how much money does she receive from the bank? How much does she repay the bank at the end of the 150 days?

51. Jane Barker needs to have $430 in order to pay her income tax. What sum should Mrs. Barker arrange to borrow from the bank if the bank charges interest in advance at 7% and the loan is to be repaid in 90 days?

6 | Ratio

Ratio is used in comparing quantities. We can compare two quantities in different ways. Thus, if we wish to compare 9 feet with 3 feet, we can subtract:

$$9 \text{ feet} - 3 \text{ feet} = 6 \text{ feet}$$

This means 9 feet is 6 feet more than 3 feet.

We are comparing two quantities by *subtraction* when we describe one quantity as a certain amount *greater* than or less than another.

When we say "Alex is 10 pounds heavier than Al" we are comparing by *subtraction*.

We also compare quantities by division. A *comparison by division* is the *ratio* of the two quantities compared. Thus when we compare the quantities 10 pounds and 5 pounds, by division, we are finding the ratio between them. The ratio of 5 pounds to 10 pounds is

$$\frac{5 \text{ pounds}}{10 \text{ pounds}} = \frac{1}{2}$$

The fraction is always reduced to lowest terms and is expressed in abstract terms. This means no units are expressed for a ratio in simplest form. If we are trying to express the ratio of 6 inches to 6 feet, we do not have $\frac{6}{6} = 1$, because one of the units is inches and the other is feet. If we express both as inches, we have

$$\frac{6 \text{ inches}}{72 \text{ inches}} = \frac{1}{12}$$

If we express both as feet, we have

$$\frac{\frac{1}{2} \text{ foot}}{6 \text{ feet}} = \frac{1}{2} \times \frac{1}{6} = \frac{1}{12}$$

Always be sure that like units are compared in writing ratios. If we are asked to find the ratio between 6 inches and 8 gallons, we can say there is *no ratio* because these units cannot be changed to like terms.

When we compare quantities by subtraction, the difference or remainder has the same denomination as the quantities themselves.

Examples

$$9 \text{ ft.} - 3 \text{ ft.} = 6 \text{ ft.}$$
$$10 \text{ lb.} - 4 \text{ lb.} = 6 \text{ lb.}$$
$$25 \text{ min.} - 10 \text{ min.} = 15 \text{ min.}$$

When we compare two quantities by division (this is the same as finding the ratio between them), the answer has no denomination:

Examples

$$\frac{8 \text{ ft.}}{12 \text{ ft.}} = \frac{2}{3}$$

$$\frac{12 \text{ lb.}}{9 \text{ lb.}} = \frac{4}{3}$$

$$\frac{18 \text{ min.}}{36 \text{ min.}} = \frac{1}{2}$$

As you continue your study of mathematics, you will find that the idea of ratio continues to be very important. It is basic in the study of calculus, for example.

EXERCISE 6

1. What is the ratio of $3 to $12?
2. What is the ratio of 12 inches to 36 inches?
3. What is the ratio of 16 pounds to 24 pounds?
4. What is the ratio of 14 feet to 26 feet?
5. What is the ratio of 6 feet to 5 yards?
6. What is the ratio of 8 inches to 2 feet?
7. What is the ratio of 48 pounds to 3 pounds?
8. What is the ratio of 1 foot to 1 mile?
9. What is the ratio of 2640 feet to $\frac{3}{4}$ mile?
10. What is the ratio of 4 pints to 8 quarts?
11. A college has 9648 students, 4326 of whom are women. What is the ratio of men students to women students?
12. What is the ratio of 4 inches to 5 feet?
13. What is the ratio of 12 ounces to 8 pounds?
14. What is the ratio of 3 yards to 15 feet?
15. What is the ratio of 40 minutes to 6 hours?
16. What is the ratio of 15 centimeters to 5 meters?
17. What is the ratio of 25 centimeters to 45 meters?
18. What is the ratio of 16 grams to 64 kilograms?
19. What is the ratio of 85 grams to 45 kilograms?
20. A baseball team wins 5 games out of every 11 games it plays. Find the ratio of the games it wins to the games it plays.
21. In the above problem, find the ratio of the games won to the games lost.
22. Find the ratio of 60 centimeters to 80 meters.
23. James hits a target 5 out of every 15 shots. What is the ratio of his hits to his misses?
24. What is the ratio of 8 quarts to 4 gallons?

25. What is the ratio of 6 pounds to 84 ounces?

26. What is the ratio of 4 miles to 8360 yards?

27. One circle has a diameter of 15 inches, and another circle has a diameter of 9 inches. What is the ratio of the radius of the first circle to the radius of the second?

28. In drawing plans for a house assume that $\frac{1}{4}$ inch represents 1 foot of actual length of the house. What is the ratio between the distance on the plan and the actual distance on the ground?

29. On a certain map one inch represents 20 miles. What is the ratio of the distances on the map to the actual distances on the earth?

7 | Proportion

A proportion is a statement that two ratios are equal. Thus $\frac{4}{8} = \frac{1}{2}$ is a proportion that may also be written in the form $4:8 = 1:2$. The statement $\frac{4}{8} = \frac{1}{2}$ is read, "four over eight equals one over two," or "four-eighths equals one-half." The statements $\frac{4}{8} = \frac{1}{2}$ and $4:8 = 1:2$ may both be read, "four is to eight as one is to two."

Each ratio involved in a proportion may refer to the same denomination as in the next example in which both ratios refer to feet.

$$4 \text{ ft.} : 8 \text{ ft.} = 9 \text{ ft.} : 18 \text{ ft.}$$

However, the ratios may refer to different denominations, such as:

$$9 \text{ ft.} : 12 \text{ ft.} = 15 \text{ lb.} : 20 \text{ lb.}$$

The two ratios can be expressed as a proportion, since each ratio equals $\frac{3}{4}$.

A proportion contains four terms, since the proportion is made up of two ratios and each ratio has two terms. Again we see that proportions can be written as fractions:

$$\frac{4 \text{ ft.}}{8 \text{ ft.}} = \frac{9 \text{ ft.}}{18 \text{ ft.}}$$

$$\frac{9 \text{ lb.}}{12 \text{ lb.}} = \frac{15 \text{ lb.}}{20 \text{ lb.}}$$

The four terms of a proportion are usually described as *first term*, *second term, third term,* and *fourth term.* Thus in the proportion $\frac{4 \text{ ft.}}{8 \text{ ft.}} = \frac{9 \text{ ft.}}{18 \text{ ft.}}$, 4 ft. is the first term, 8 ft. the second, 9 ft. the third, and 18 ft. the fourth term.

The first and fourth terms are known as the *extremes* of the proportion. The second and third terms are called the *means*.

The fourth term is also referred to as the *fourth proportional* to the first three terms.

We have seen that a proportion is an equation stating that two ratios are equal. Therefore we may solve a proportion as we solve any equation.

The proportion $a:b = c:d$ may be written in fractional form $\frac{a}{b} = \frac{c}{d}$.

Multiplying both sides of the equation by bd, we have

$$ad = bc$$

We see that the term ad is the *product of the extremes* of the proportion and the term bc is the *product of the means.* The following important statement concerning a proportion results.

In any proportion the product of the means equals the product of the extremes.

Using this principle we can find any missing term in a proportion. For example,

$$3:x = 9:16$$

The product of the means is $9x$. The product of the extremes is 48. Now we can write

$$9x = 48$$

This gives $x = 5\frac{1}{3}$.

There are proportions where the second term is the same as the third term. Consider the proportion:

$$2:8 = 8:32$$

In this example, the second and third terms are each 8. When the second and third terms are the same quantity, this quantity is known as *the mean proportional between the other two terms.* In this case there are only three *different* quantities and the fourth term is known as the *third proportional* to the first and second terms. In the example

$$2:8 = 8:32,$$

ELEMENTARY ALGEBRA 175

32 is the third proportional to 2 and 8. Eight is the mean proportional between 2 and 32.

Consider the proportion
$$a:b = b:c$$

Here, b is the mean proportional between a and c. From this equation we find
$$b^2 = ac$$
or
$$b = \pm\sqrt{ac}$$

Thus a convenient way to find the mean proportional between 3 and 48 is as follows.

Set up the proportion
$$3:x = x:48$$
Now
$$x^2 = 144$$
and
$$x = \pm 12$$

The mean proportional may be either 12 or -12.

In checking the solution, we find
$$3:(+12) = (+12):48$$
or
$$\frac{3}{12} = \frac{12}{48}$$
Then
$$\frac{1}{4} = \frac{1}{4}$$

Checking the other value -12,
$$3:(-12) = (-12):48$$
or
$$\frac{3}{-12} = \frac{-12}{48}$$
Then
$$\frac{-1}{4} = \frac{-1}{4}$$

Proportion has many practical uses.

For example, if a car can go 96.3 miles on 6.2 gallons of gasoline, how many miles can it go on 12 gallons of gasoline? Following the rules for approximate numbers, after setting up the proportion, we have $\frac{96.3}{x} = \frac{6.2}{12}$. In this proportion, x represents the number of miles the car can travel on 12 gallons. Simplifying by multiplying both sides of the equation by $12x$, we have

$$12x\left(\frac{96.3}{x}\right) = 12x\left(\frac{6.2}{12}\right)$$

Then:

$6.2x = 12(96.3)$

$6.2x = 1155.6$ or 1160

$x = 187$ or 190 miles, rounding to two significant figures

EXERCISE 7

Solve for the unknown in each of the following proportions:

1. $3:x = 4:48$
2. $8:10 = x:90$
3. $x:5 = 15:45$
4. $5:4.5 = 9:y$
5. $6:x = x:54$
6. $y:4 = 16:y$
7. $8:(3+x) = 4:13$
8. $8:(k-2) = 24:(4k+1)$
9. $7:(k-2) = 21:(4k+1)$
10. $560:484 = 9:x$
11. $5.8:8 = 29.0:y$
12. $1:4.3 = 6.7:k$

13. Find the fourth proportional to the three numbers 3, 4, and 5.
14. Find the mean proportional between 8 and 16.
15. Find the mean proportional between 4 and 49.
16. What is the mean proportional between 27 and $\frac{1}{3}$?
17. A recipe calls for $1\frac{1}{2}$ cups of flour. The recipe is planned for ten people. How many cups of flour are needed in planning for fifteen people?
18. In any elastic body the distortion is proportional to the distorting force if kept within what is referred to as the "elastic limit." This is Hooke's law. If a force of 6 pounds will stretch the spring 5 inches, what force will stretch it 10 inches?
19. If 1 kilogram (2.2 pounds) stretches a spring $1\frac{3}{4}$ inches, what force is required to stretch it 1 foot 9 inches at the same rate?

ELEMENTARY ALGEBRA   177

20. If a car can go 108 miles on 7.8 gallons of gasoline, how far can it go on 46 gallons of gasoline?

21. A car travels 103 feet per second. Find its speed in miles per hour.

22. A conveyor belt carries 28 cubic feet of coal per minute. How many cubic yards will be carried in 8 hours?

23. Change 1260 gallons per hour into quarts per second.

24. A 100-pound bag of fertilizer contains 22 pounds of limestone. How many pounds of limestone will 86.7 pounds of fertilizer contain?

25. If 46 identical bolts weigh 4.6 pounds, how many pounds will 28 of these bolts weigh?

26. If 8.0 pounds of pecans sell for $5.60, how much will the selling price be for 10 pounds?

27. Using Problem 26, how many pounds of pecans can be bought for $120?

28. Dress material is priced at 3.0 yards for $2.49. How much will $8\frac{1}{2}$ yards of the material cost?

29. A fertilizer mixture contains 2 parts nitrogen, 2 parts potash, and 3 parts phosphate by weight. How many pounds of the mixture will contain 60 pounds of potash?

30. A paving mixture contains 2 parts cement, 2 parts sand, and 3 parts gravel by weight. How many pounds of the mixture can be made with 3258 pounds of cement?

31. The volumes of similar solids are proportional to the cubes of corresponding dimensions. If the diagonal of a cube is doubled, what change is made in the volume?

32. The year's profits in the B and C Company were $32,000. Mr. Smith has $16,000, Mr. Jones $28,000, and Mr. Baker $24,000 invested in the business. If the profits were distributed according to each partner's investment, how much did each receive?

33. Two boys divide 40 marbles in the ratio of 3:1. How many marbles does each boy have?

34. Two girls divide $14.80 in the ratio of 5:3. How much does each girl receive?

Chapter 9

Measurement of Plane and Solid Figures

1 | Perimeter and circumference formulas

The words perimeter and circumference both refer to distance around. However, perimeter refers to the distance around a closed broken line in a plane. Such a figure has examples in the triangle, square, and rectangle. A *triangle* is a figure formed in a plane by connecting three points that are not in a straight line. A *square* is a closed four-sided figure in a plane. A square has equal sides and equal angles. A *rectangle* is a closed four-sided figure in a plane. All the angles of a rectangle are right angles and all the angles of a square are right angles. How then, does a rectangle differ from a square? The triangle, square, and rectangle are illustrated in Figure 6.

Circumference refers to the distance around a circle or the length of the curved line which is the circle. While we could speak of the perimeter of a circle this generally is not done.

The perimeter of a closed broken line in a plane is the sum of the lengths of its sides. Thus if we let p stand for perimeter and let a, b, and c refer to the sides of a triangle, then $p = a + b + c$. (See Figure 7.)

For example, find the perimeter of a triangle if the sides are 12.4 feet, 6.8 feet, and 14.5 feet. Assuming that all measurements are approximate, we have $p = 12.4 + 6.8 + 14.5$ or 33.7 feet.

Triangle
(a)

Square
(b)

Rectangle
(c)

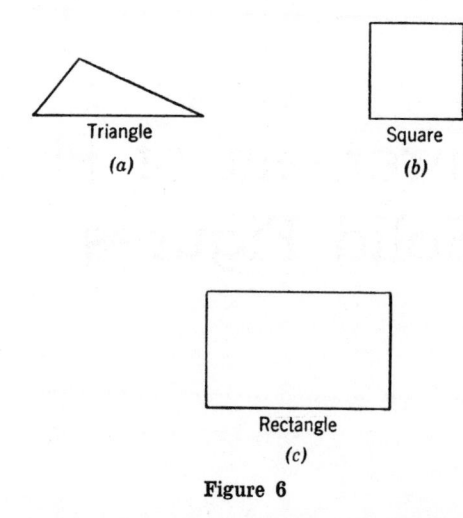

Figure 6

You may work out your own set of experiments to determine how many times the diameter of a circle is contained in the circumference. The objects used should be sufficiently large to give reasonably correct answers. Experiments with larger and smaller objects generally lead to the conclusion that better results are obtained if the object used is not too small and is carefully made. A circular table, earthen pipe, large crock, dinner plate, pan, or automobile tire, all furnish ample material for this kind of work. Through these experiments, the length of the diameter and circumference are obtained, and you will find that, regardless of the size of the circle used, the quotient of the circumference divided by the diameter is approximately the value 3.1416. This number is represented by the Greek letter π (read pi). It has been mathematically proved that the value of π cannot be expressed exactly as a whole number or as a combination of whole number and fraction.

The circumference of a circle equals two times pi times the radius, or $2\pi r$. Since the radius of a circle is half the diameter, the circumference also may be written πd. Let us use π as 3.1416. Of

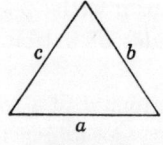

Figure 7

MEASUREMENT OF PLANE AND SOLID FIGURES — 181

course this is an approximate number. Diameter and radius measurements should also be considered as approximate numbers.

For example, find the circumference of a circle whose radius is 8.75 inches. Now $C = 2\pi r$ or $C = 2(3.142)(8.75) = 55.0$ inches.

EXERCISE 1

Find the perimeters of the following figures.

1. A triangle whose sides are 32.7 yards, 21.9 yards, and 14.8 yards.

2. A rectangle two of whose sides are 168.3 feet and 79.4 feet.

3. A triangle with one side 18.4 rods and the two other sides each equal to 27.6 rods.

4. A square whose side is 41.3 rods.

5. If l represents the length of a rectangle and w its width, write a formula for its perimeter, p.

6. Find the perimeter of a rectangle two of whose sides are 64.35 feet and 23.7 feet.

7. Find the circumference of a circle if its radius is 38.64 feet.

8. If s represents the side of a square, write a formula for its perimeter, p.

9. Find the circumference of a circle if its diameter is 247.8 feet.

10. Find the circumference of a circle if its radius is 54.6 rods.

11. Find the perimeter of a rectangle if its length is 4.68 and its width is 3.4.

12. Find the circumference of a circle if its diameter is $33\frac{3}{4}$ feet.

13. What is the perimeter, p, of a semicircle whose radius is r?

14. What is the circumference of a circle whose diameter is 9.64 inches?

15. The length of a rectangle is $4\frac{1}{2}$ times its width, and its perimeter is 68 feet. What are its dimensions?

16. The perimeter of a rectangle is 140 feet, and its length is three times the width. What are its dimensions?

17. The length of a rectangle is 4 less than five times its width, and the perimeter is 64. Find its dimensions.

18. Two sides of a triangle are equal in length, and the third side is 12 more than the length of the equal sides. If the perimeter of the triangle is 156, what are its sides?

19. The length of a rectangle is 40% longer than its width. If the perimeter of the rectangle is 528 inches, what are its dimensions?

20. A side of the larger of two squares is five times a side of the smaller, and the sum of the perimeters of the two squares is 264. Find the sides of the two squares.

2 | Area formulas for the rectangle, parallelogram, and triangle

The formula for the area of the rectangle is based on direct measurement. In measuring this area, let us use a square as a

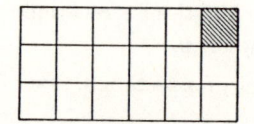

Figure 8

convenient unit of measure. To measure the area of the rectangle we find the number of times the rectangle contains the square. Consider a rectangle 6 centimeters long and 3 centimeters wide, and assume the unit of measure to be a square centimeter as in Figure 8. We see that the square is contained in the given rectangle 18 times. This fact is obtained very quickly by observing that in each row there are 6 of the square centimeters. Since there are 3 rows, the total number of square centimeters in the given rectangle is 6 × 3, or 18 square centimeters. We notice that the number of squares in each row is exactly equal to the number of linear units in the length of the rectangle, and the number of rows is equal to the number of linear units in the width of the rectangle. The rule for finding the area of a rectangle is as follows: The area of a rectangle is equal to the product of the number of units of length and the number of units of width. The result is units of area, or square units. For practical purposes we find it more convenient in computing areas of rectangles to use the formula $A = L \times W$. This statement says in words, of course,

MEASUREMENT OF PLANE AND SOLID FIGURES 183

that the area of a rectangle is equal to the product of the length and the width.

For example, find the area of a rectangle if its length is 8 feet and its width is 5 feet. In this illustration, $L = 8$ feet and $W = 5$ feet. Then $A = 8 \times 5$, or 40 square feet. As we have commented before, it is very important that units be placed in the answer in a form consistent with the way they appear in the given problem. Thus in the illustration above, feet are given in the problem and the answer is expressed in square feet.

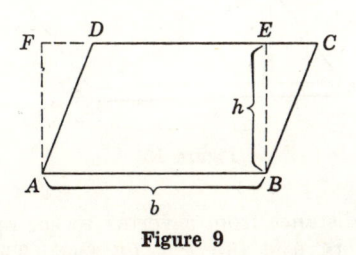

Figure 9

Consider now the parallelogram $ABCD$ in Figure 9. A *parallelogram* is a closed four-sided figure in a plane; its opposite sides are *parallel*. *Parallel* lines are lines that are equidistant apart. In Figure 9, AB and DC are parallel lines, and AD and BC are parallel lines. The line AB on which the parallelogram appears to be standing is called the base of the parallelogram. The shortest distance BE from the base AB to the opposite side DC is called the height of the parallelogram. If we take away the right triangle* BEC and consider it as placed on the left where triangle AFD is shown, a rectangle is made having the same base, height, and area as the original parallelogram. Hence, the number of units of area in a parallelogram is equal to the product of the number of units of length in the base and the number of units of length in the height. Expressing this idea simply as a formula, we have $A = b \times h$.

As an illustration, find the area of a parallelogram, the base of which is 18 inches and the height 5 inches. Here, $b = 18$ inches and $h = 5$ inches. Then $A = 18$ inches $\times$ 5 inches $= 90$ square inches.

Turning now to the finding of the area of a triangle, let us examine

---

* A right triangle is a triangle that contains a right angle. A right angle equals 90 degrees.

triangle ABC in Figure 10. We assume that ABC is a completely general triangle. If another triangle exactly the same size as ABC is located as indicated by dotted lines in Figure 10, a parallelogram appears. We see then that the area of a triangle is one-half the area of the parallelogram having the same base and height. Thus $A = \frac{1}{2} \times b \times h$. Now, find the area of a triangle whose base is 18 feet and whose height or altitude is 4.5 feet. An altitude of a triangle is

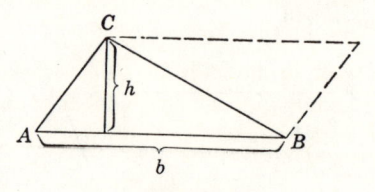

Figure 10

the perpendicular distance from a vertex to the opposite side. In this example, $b = 18$ feet and $h = 4.5$ feet. Then $A = \frac{1}{2} \times 18$ feet $\times$ 4.5 feet, or 40.5 square feet. However, remembering that measurements are approximate numbers and following the rules explained in Chapter Seven on approximate numbers, we round this result to 40 square feet.

EXERCISE 2

1. Find the area of a rectangle that is 9.8 inches long and 5.3 inches wide.

2. Find the area of a rectangle having the dimensions 18.6 rods by 9.7 rods.

3. Find the area of a triangle if its base is 64 inches and its height is 22 inches.

4. A triangle has a base 88 feet long and an altitude 66 feet. Find its area.

5. A parallelogram has a base of 36.8 feet and an altitude of 16.9 feet. Find its area.

6. Find the area of a parallelogram if its base is 98 inches and its height is 54 inches.

MEASUREMENT OF PLANE AND SOLID FIGURES _____185

7. A rectangular grove is 1865 feet long and 1428 feet wide. How many acres does the field contain?

8. Find the area of a rectangle 48 inches by 5.2 feet.

9. How many bricks will be used in paving a street 35 feet wide and 8.3 miles long? Assume that $7\frac{1}{2}$ bricks are needed per square foot.

10. A piece of land one mile square is known as a section. How many acres are there in a section?

11. Laths are sold in bundles of 100. They are 4 feet long, approximately $1\frac{3}{8}$ inches wide, and are generally nailed about $\frac{3}{8}$ of an inch apart. How many square yards of wall can be covered with a bundle?

12. Assuming that in planning a schoolroom, 15 square feet are used as the recommended floor space for each pupil, how many students can be housed in a room 22 feet by 36 feet?

13. A room is 18 feet wide and 26 feet long. How much will it cost to cover the floor of the room with linoleum if the linoleum costs $5 per square yard?

14. Find the number of bundles of laths that will be needed to lath the walls and ceiling of a room 25 feet long, 18 feet wide, and 9 feet high. (See Problem 11.)

15. If two parallelograms have the same base, 18 feet, and have altitudes of 15 feet and 10 feet, respectively, how do their areas compare?

16. A room is 18 feet long and 14 feet wide. A rug is on the floor and the dimensions of this rug are 16 by 12 feet. How many square feet of floor space are not covered by the rug?

17. The flight deck of a carrier is 9.03 times as long as it is wide. Find its width if its length is 948 feet.

18. Find the area in square yards of the flight deck in the problem above.

19. One triangle has a base of 44 inches and a height of 20 inches. A second triangle has a base of 11 inches. How high should it be in order to have an area three times that of the first triangle?

20. Aircraft plywood comes in sheets 48 by 72 inches. Find the price per square foot if it sells for $10.50 per sheet.

3 | Area formula for the trapezoid

The rule for finding the area of a trapezoid comes easily from a study of Figure 11. Let $ABCD$ represent any trapezoid with a and b the lengths of the parallel sides. The area of the trapezoid is

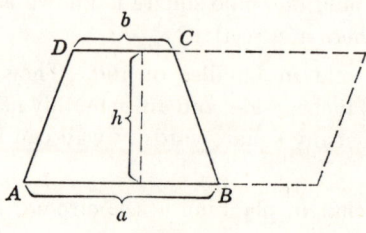

Figure 11

one-half the area of a parallelogram having the same height, but its base is equal to the sum of the lengths of the parallel sides of the trapezoid. Written as a formula, this is $A = \frac{1}{2}h(a + b)$.

For example, find the area of a trapezoid whose bases are 22 and 12 inches respectively, and whose height is 15 inches. Here we have $a = 22$ inches, $b = 12$ inches, and $h = 15$ inches. Then:

$$A = \frac{1}{2}(15)(22 + 12) \text{ or } A = \frac{1}{2}(15)(34)$$

Finally, $A = 260$ square inches, rounding to two significant figures.

EXERCISE 3

Find the areas of the trapezoids in Problems 1 through 10.

1. The bases are 10 and 7, and the altitude is 8.
2. The bases are 30 and 16, and the altitude is 12.
3. The bases are 484 feet and 126 feet, and the altitude is 88 feet.
4. The bases are 84 yards and 32 yards, and the altitude is 66 feet.
5. The bases are 11.4 centimeters and 7.6 centimeters, and the altitude is 5.9 centimeters.

MEASUREMENT OF PLANE AND SOLID FIGURES _____187

6. The bases are 39.8 meters and 17.5 meters, and the altitude is 12.6 meters.

7. The bases are 147.9 kilometers and 92.8 kilometers, and the altitude is 81.7 meters.

8. The bases are 120.3 rods and 186.9 rods, and the altitude is 92.3 rods.

9. The bases are $138\frac{1}{2}$ inches and 19 feet, and the altitude is 84 inches.

10. The bases are 34.6 miles and 23.7 miles, and the altitude is 4840 feet.

11. In trapezoid $ABCD$ draw diagonal AC dividing the trapezoid into two triangles. Using the formula for finding the area of a triangle, develop the formula for finding the area of a trapezoid.

12. Find the formula for the area of Figure 12.

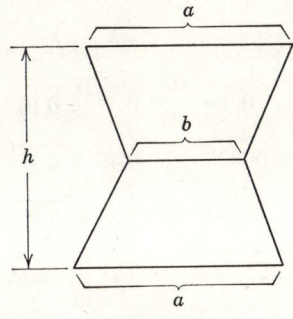

Figure 12

13. Find this area if $a = 45$ inches, $b = 25$ inches, and $h = 46$ inches.

14. The area of a trapezoid is 796 square feet. Its parallel bases are 238 feet and 165 feet respectively. What is the height of the trapezoid?

15. Sheldon Park, which has the shape of a trapezoid, has two parallel sides 440 feet and 368 feet long, and is 264 feet deep. How many acres does it contain?

4 | Square root

Let us think first about finding the squares of numbers. Two squared, or 2×2, or 2^2 equals 4. Five squared, or 5×5, or 5^2 equals

25. In finding the square root of a number we are actually determining what smaller value, when multiplied by itself, will equal this number. Thus 6 is the square root of 36, and 12 is the square root of 144.

Let us consider the following examples:

$$(a)$$

$$1^2 = \widehat{1}$$
$$2^2 = \widehat{4}$$
$$5^2 = \widehat{25}$$
$$7^2 = \widehat{49}$$
$$9^2 = \widehat{81}$$

$$(b)$$

$$10^2 = \widehat{100}$$
$$14^2 = \widehat{196}$$
$$23^2 = \widehat{529}$$
$$76^2 = \widehat{5776}$$
$$94^2 = \widehat{8836}$$

$$(c)$$

$$100^2 = \widehat{10000}$$
$$225^2 = \widehat{50625}$$
$$669^2 = \widehat{447561}$$
$$853^2 = \widehat{727609}$$
$$966^2 = \widehat{933156}$$

$$(d)$$

$$0.1^2 = 0.\widehat{01}$$
$$0.4^2 = 0.\widehat{16}$$
$$2.3^2 = \widehat{5.29}$$
$$3.7^2 = \widehat{13.69}$$
$$10.29^2 = \widehat{105.8841}$$

In column a we observe that the square root of a one- or two-digit number is in units, or is one place to the left of the decimal point. In column b we see that the square root of a three- or four-digit number is in tens, or two places to the left of the decimal point. In column c we find that the square root of a five- or six-digit number is in hundreds, or three places to the left of the decimal point. In column d we see that the square root of numbers one or two places to the right of the decimal is in tenths; the square root of numbers three or four places to the right of the decimal is in hundredths, and so forth.

In taking the square root of a number we pair off the digits to the left and the right of the decimal point in the number the square root of which is to be found. For example, to find the square root of

MEASUREMENT OF PLANE AND SOLID FIGURES _____189

4, 196, 1936, 95481, 6342.7, and 85,4362 we have

$4 = \widehat{4}$; $\sqrt{4}$ will be in one digit

$196 = \widehat{196}$; $\sqrt{196}$ will be in two digits

$1936 = \widehat{19}\widehat{36}$ $\sqrt{1936}$ will be in two digits

$95481 = \widehat{9}\widehat{54}\widehat{81}$; $\sqrt{95481}$ will be in three digits

$6342.7 = \widehat{63}\widehat{42}.\widehat{7}$; $\sqrt{6342.7}$ will be in two digits and one decimal place

$85.4362 = \widehat{85}.\widehat{43}\widehat{62}$; $\sqrt{85.4362}$ will be in one digit and two decimal places

For example, find the square root of 3136.

1. Point off the digits in pairs, placing the decimal point above the line immediately above its position in the problem. $\sqrt{\widehat{31}\widehat{36}}$.

2. Find the largest square contained in 31, the first grouping. Here $5 \times 5 = 25$, is the largest perfect square contained in 31. The number is 5; place it in the answer just above the 1 in 31.

$$\begin{array}{r} 5\ 6. \\ \sqrt{3136.} \\ 25 \end{array}$$

100	636
6	
106	636

3. Square the 5, which is the only number in the answer thus far, obtaining 25. Subtract this from 31. This gives 6.

4. Bring down the next group of two digits, 36.

5. Double the answer thus far obtained, $2 \times 5 = 10$. Annex one zero to the 10, making it 100. Use this as a trial divisor. Thus 636 divided by 100 will give approximately 6. Add 6 to the trial divisor; 100 plus 6 equals 106. Place 6 in the answer just above the 6 in 36. Multiply the 106 by 6. This gives 636.

6. Subtract 636. Since there is no remainder, 3136 is the perfect square of 56. If the number 3136 were not a perfect square, we would need to annex two zeros at a time and continue the process of solution until the desired degree of accuracy had been reached.

In examining square roots of approximate numbers we make an observation. The square of the approximate number 16.8, when rounded according to rules for multiplying approximate numbers, is 282. In taking the square root of 282 we find 16.8. *The square root of a number contains the same number of significant digits as the given number.* Thus to find the square root of an approximate number with x number of significant digits, carry the answer to $x + 1$ significant digits, and round the answer one place.

As another example, find the square root of 7.34, and assume it to be an approximate number.

$$\begin{array}{r} 2.\,7\,\;0\,\;9 \\ \sqrt{7.340000} \end{array}$$ which rounds to 2.71 for the answer

```
              2. 7 0 9
           √‾7.340000‾
              4
      40  |  3 34
       7  |
      ——  |
      47  |  3 29
     5400 |  50000
        9 |
      ——  |
     5409 |  48681
```

5 | The Pythagorean theorem and some of its applications

In a right triangle the side opposite the right angle is the *hypotenuse*, whereas the other two sides are the *legs*. These names are not given to the sides of any other kind of triangle.

The right triangle is conventionally labeled as shown in Figure 13. The capital letters A, B, and C designate the angles, C always indicating the right angle. The small (lower case) letters a, b, c

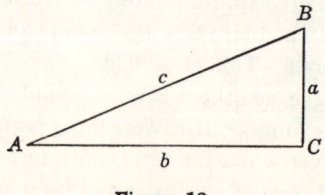

Figure 13

represent the sides opposite the angles A, B, and C, respectively. Of course c is always the hypotenuse; a, the leg opposite angle A; and b, the leg opposite angle B. With respect to angle A, a is known as the opposite leg and b, the leg adjacent. Of course the adjectives are reversed with respect to angle B, since b stands for the opposite leg and a for the adjacent leg.

The distinctive property of the right triangle is expressed algebraically in the equation $c^2 = a^2 + b^2$ and by the equivalent equations $a^2 = c^2 - b^2$ and $b^2 = c^2 - a^2$. This property is expressed geometrically in the famous theorem of Pythagoras, which states that the area of the square whose side is c equals the sum of the areas of the squares whose sides are a and b.

Example 1

If the sides of a rectangle are 10 feet and 12 feet, how long is a diagonal of the rectangle? In many instances a figure is desirable in understanding a problem.

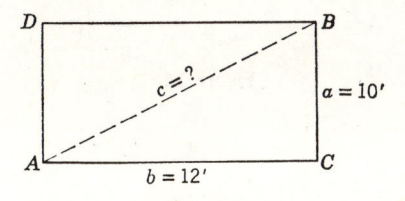

Figure 14

In Figure 14 we have rectangle $ACBD$ with diagonal c and right triangle ACB with hypotenuse c and sides a and b. From the Pythagorean theorem we have $c^2 = a^2 + b^2$.

Substituting from the figure we have

$$c^2 = (10)^2 + (12)^2,$$

or

$c = \sqrt{100 + 144}$

$c = \sqrt{244}$ or 15.6 feet

$c = 16$ feet, rounding according to approximate numbers

Example 2

If a ladder 12 feet long rests against a wall, with the foot of the ladder 7 feet from the foot of the wall, how high up on the wall is the top of the ladder?

We can clear this problem up rather nicely with the help of a diagram.

In Figure 15 we see that $c = 12$ feet and $b = 7$ feet. But $c^2 = a^2 + b^2$, and $a^2 = c^2 - b^2$. Then

$$a^2 = (12)^2 - (7)^2,$$

or

$$a = \sqrt{144 - 49}$$
$$a = \sqrt{95} = 9.75 \text{ or } 10 \text{ feet where the zero is not significant.}$$

This tells how high up on the wall the ladder reaches.

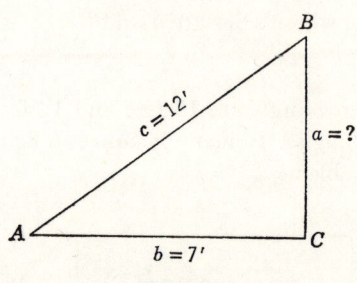

Figure 15

EXERCISE 4

Find the square roots of the following 30 problems, assuming that the numbers given in each case are approximate numbers. The answers should be rounded accordingly.

1. 64
2. 36
3. 144
4. 196
5. 2601
6. 3844
7. 7225
8. 8649
9. 11,236
10. 11,881
11. 13,689
12. 15,129
13. 17,956
14. 19,881
15. 24,025
16. 28,561
17. 1288.81
18. 1722.25
19. 9139.36
20. 7656.25
21. 10,506.25
22. 11,004.01
23. 896,478.3
24. 208,496.7
25. 0.0073
26. 0.00081
27. 0.02783
28. 0.09643

MEASUREMENT OF PLANE AND SOLID FIGURES   193

29. Find the side of a square whose area is 441 square feet.

30. Find the side of a square whose area is 576 square centimeters.

31. The sides of a right triangle are 15 feet and 19 feet. Find the hypotenuse.

32. A rope 98 feet long is attached to the top of a flagpole and reaches to a point on the ground 72 feet from the foot of the pole. How high is the pole?

33. If a tether ball swings on a rope 11 feet long, from the top of a pole 13 feet high, how far from the pole is the ball when it is at a level 4 feet above the ground?

34. Find the length of a side of a square whose diagonal is 12 feet longer than a side.

35. Find the altitude of an isosceles triangle if the equal sides are 10 inches and the base is 12 inches. (An isosceles triangle is a triangle with two equal sides.)

36. Find the length of a rafter for a roof if the height of the roof is 12 feet and its span is 34 feet.

37. During a storm, a tree 54 feet high snaps so that the upper portion, which still remains attached to the trunk where it snapped, touches the ground 32 feet from the foot of the tree. How high up is the break in the tree?

38. Find the length of the diagonal of a rectangular box having dimensions of 8 feet, 6 feet, and 4 feet.

6 | Area formulas for circle and ellipse

A circle is a closed curve every point of which is equally distant from a point within, called the center. If r is the radius of a given circle, its area is given by the formula $A = \pi r^2$. Here π is a number which can never be expressed exactly as a common or decimal fraction. It is equal to $\frac{22}{7}$ or 3.14159, both of which are approximate values.

For example, find the area of a circle whose radius is 23.5 inches. Let us use π here as 3.14, carrying the same number of digits as we have in the radius. Now since $A = \pi r^2$, $A = 3.14(23.5)^2$, or $A = 3.14(552.2)$, when we round the $(23.5)^2$ to four significant figures (one more than we expect to use in the answer). Then $A = 1730$ square inches.

The ellipse is a beautiful figure rather frequently used in designs. Elliptical gears and cams are frequently used in machinery. Elliptical arches often appear in bridges and on the paneling in furniture. Furthermore, each planet moves in an ellipse, although until the time of the astronomer Kepler (1571–1630) it was thought that the planets moved in circles.

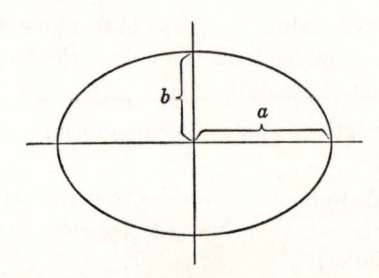

Figure 16

In the ellipse, the long diameter is called the *major axis* and the short diameter is called the *minor axis*. In Figure 16 an ellipse is shown; a is called the semi-major axis and b, the semi-minor axis.

The area of an ellipse is πab or π times the semi-major axis times the semi-minor axis. This formula is $A = \pi ab$.

For example, find the area of the ellipse whose semimajor axis is 24.2 inches and whose semiminor axis is 16.5 inches. Let us use 3.14 for π, keeping a consistent number of significant figures. Now $A = (3.14)(24.2)(16.5)$, or $A = (3.14)(399.3)$. Then $A = 1250$ square inches. Here the zero is not significant, of course.

EXERCISE 5

Find the areas of the following circles. In each case round π to a consistent number of significant figures, using $\pi = 3.14159$ as an approximate value from which to start the rounding.

1. Radius = 8 inches
2. Radius = 12 inches

MEASUREMENT OF PLANE AND SOLID FIGURES 195

3. Radius = 20 inches
4. Radius = 14 inches
5. Radius = 28.4 meters
6. Radius = 46.5 centimeters
7. Radius = 124.8 centimeters
8. Radius = 96.2 meters
9. Radius = 85.7 feet
10. Radius = 145.9 rods
11. Radius = 28.6 rods
12. Radius = 264.4 feet
13. Diameter = 64 inches
14. Diameter = 48 centimeters
15. Diameter = 76 centimeters
16. Diameter = 85 inches
17. Diameter = 168 kilometers
18. Diameter = 0.85 mile
19. Diameter = 0.42 mile
20. Diameter = 146 kilometers

Find the areas of the following ellipses, rounding as was instructed for the circles.

21. Semimajor axis 14; semiminor axis 10
22. Semimajor axis 48; semiminor axis 25
23. Semimajor axis 52 inches; semiminor axis 26 inches
24. Semimajor axis 84 feet; semiminor axis 46 feet
25. Semiminor axis 184 kilometers; semimajor axis 468 kilometers
26. Semiminor axis 36 rods; semimajor axis 149 rods
27. Semimajor axis 15 feet; semiminor axis 80 inches
28. Semimajor axis 24 feet; semiminor axis 168 inches
29. Semimajor axis 5674 meters; semiminor axis 3465 centimeters
30. Semimajor axis 2386 meters; semiminor axis 1263 centimeters

31. Find the area of the largest circle that can be cut from a 20-inch square.

32. In Problem 31, the area of the circle is what per cent of the area of the square?

33. Find the cost of a sidewalk 8 feet wide around a circle whose diameter is 168 feet, if the cost is 64 cents a square foot.

34. Find the formula for the area of a circular ring if r_1 is the radius of the outside circle and r_2 is the radius of the inside circle.

35. What is the radius of a circular tract that contains three acres?

36. Solve the formula $A = \pi r^2$ for r.

37. What is the radius of a circle that has an area equal to that of an 18-foot square?

7 | Rectangular solid and pyramid

Geometric figures with three dimensions are called solids. These figures have length, width, and height or depth.

A rectangular solid is a figure that is formed by six rectangles that are parallel to each other in pairs. If the rectangles are squares, the rectangular solid formed is a cube.

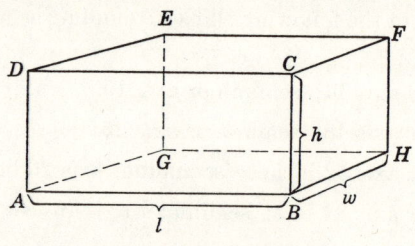

Figure 17

Figure 17 shows a rectangular solid. The rectangles $ABCD$, $BHFC$, $HFEG$, $AGED$, $ABHG$, and $DCFE$ form the solid figure. If we let $l =$ the length of the rectangular solid, and let w and h be its width and height respectively, we can rather readily see the formula for the total surface area of the rectangular solid. The area of rectangle $ABCD$ equals lh. The area of rectangle $BHFC$ is hw, and the area of rectangle $ABHG$ is lw. Now, using the rectangles in all six surfaces we have area $= 2lw + 2hw + 2lh$, or $A = 2(lw + hw + lh)$, as the total surface area of a box with top and bottom.

MEASUREMENT OF PLANE AND SOLID FIGURES _____197

Let us find the total surface area of a rectangular solid whose length measures 12.3 inches, whose width measures 6.4 inches, and whose height measures 5.7 inches. Then:

$$A = 2(12.3 \times 6.4 + 5.7 \times 6.4 + 12.3 \times 5.7).$$
$$A = 2(78.72 + 36.48 + 70.11)$$
$$A = 2(78.7 + 36.5 + 70.1)$$
$$A = 2(185.3) = 2(185) \text{ or } 370 \text{ square inches}$$

In measuring the volume of a solid we choose a unit of volume. We generally choose a cube whose edge is one linear unit in length. The most common units of volume are the cubic inch, cubic foot, or cubic centimeter.

The formula for the volume of the rectangular solid is $v = l \times w \times h$, in which $l =$ length, $w =$ width, and $h =$ height.

Let us find the volume of a bin that is 6 feet long, 4 feet wide, and 3 feet deep. Here $v = $ (6 feet)(4 feet)(3 feet) $=$ (24 square feet) $\times$ (3 feet) $= 72$ cubic feet. Rounding, since we are working with approximate numbers, we obtain 70 cubic inches.

A pyramid is a solid whose base is a polygon and whose lateral surfaces are triangles with a common vertex, which is called the *vertex* of the pyramid. A polygon is a closed plane figure bounded by straight lines. A *vertex* of a triangle is the point of intersection of two sides of the triangle.

The altitude of a pyramid is the perpendicular distance from the vertex to the base. Figure 18 shows a pyramid in which $ABCD$ is

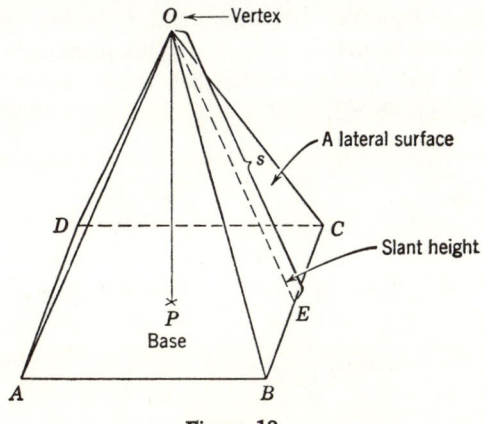

Figure 18

the base, and triangles ABO, BCO, CDO, and ADO are the lateral surfaces. OP is the altitude of the pyramid.

Of course the base of a pyramid may be any of the polygons. Those appearing most frequently as bases are the triangle, square, and rectangle.

The slant height of a pyramid is the perpendicular from the vertex to one edge of the base of a regular pyramid. Thus OE represents the slant height. In defining a *regular pyramid* we use the term *regular polygon*. A regular polygon is one all of whose sides are equal and all of whose angles are equal. If the base of a pyramid is a regular polygon and if the perpendicular from the vertex to the base passes through the center of the base, the pyramid is known as a *regular pyramid*.

The lateral surface of the pyramid refers to the sum of triangles ABO, BCO, CDO, and DAO in Figure 18. *The area of the lateral surface of a regular pyramid equals one-half the product of the perimeter of the base and the slant height.* Using the perimeter of the base as p and the slant height as s, we have A or area $= \frac{1}{2}ps$.

Find the lateral area of a regular pyramid whose slant height is 12.3 inches and the perimeter of whose base is 84.6 inches. Now $A = \frac{1}{2}(84.6)(12.3)$ or $A = (42.3)(12.3) = 520.29$ or 520 square inches.

The volume of a pyramid equals one third of the product of its altitude and its base. The formula here is $v = \frac{1}{3}Bh$, where B refers to the area of the base and h is the height.

Find the volume of the pyramid if its base is a 6-inch square and its altitude is 7 inches. Here the area of the base is 6 inches × 6 inches or 36 square inches. Then $v = \frac{1}{3}(36 \text{ square inches})(7 \text{ inches}) = 84$ cubic inches or 80 cubic inches when rounded according to approximate numbers.

EXERCISE 6

1. Find the volume of a rectangular solid $4\frac{1}{2}$ inches by 6 inches by 5 inches.

2. Find the volume of a rectangular solid $3\frac{3}{4}$ feet by 8 feet by 6 feet.

3. Find the volume of a rectangular solid 6.5 meters by 9.75 meters by 98.6 centimeters.

MEASUREMENT OF PLANE AND SOLID FIGURES ―――――199

4. Find the volume of a rectangular solid 3.4 meters by 5.8 meters by 47.3 centimeters.

5. A rectangular solid is 9 inches wide and 17 inches long and contains 1530 cubic inches. Find its height.

6. A box is to be 18 inches long and 10 inches wide. How deep must it be made to hold 900 cubic inches?

7. A rectangular gas drum has a square base 10 feet by 10 feet, and is 18 feet deep. How many thousand gallons will it hold?

8. A bin 18.5 feet long and 12.4 feet wide is filled 8.3 feet deep with maize. How many bushels of maize are in the bin? Use 2150 cubic inches as a bushel.

9. Find the cost of digging a cellar 45 feet long, 35 feet wide, and 8 feet deep at $2.20 per cubic yard.

10. A coal storage bin is 12 feet long, 9 feet wide, and 5 feet deep. Find how many tons the bin will hold if the coal weighs 52 pounds per cubic foot.

11. How many bushels of potatoes can be stored in a bin 14 feet long, 10 feet wide, and 4 feet deep? (A bushel of potatoes requires approximately 1.25 cubic feet.)

12. How many bricks are necessary to pave a patio 42 feet wide and 55 feet long? Assume that 7.5 bricks are needed per square foot.

13. Find the altitude of a pyramid whose volume is 1296 cubic inches and whose base has an area of 324 square inches.

14. A pyramid has a square base, 6 feet on each side, and an altitude of 4 feet. Find the volume of the pyramid.

15. If the base of a pyramid is a square 18 inches on a side, and the altitude is 14 inches, what is the volume?

16. Find the number of square yards of canvas necessary to make a tent in the form of a pyramid whose slant height is 14 feet and whose base is 18 feet square. Allow 4 square yards extra for seams and waste.

17. A regular pyramid has a square base whose edge is 63.4 centimeters. The slant height is 96.7 centimeters. Find the lateral area of the pyramid.

18. A regular pyramid has a square base whose edge is 14.5 feet. The slant height is 19.8 feet. Find the lateral area of the pyramid.

19. Find the total area of the pyramid in Problem 18.

20. A church steeple is in the shape of a regular pyramid whose base has 8 sides. If the slant height is 28.6 feet and an edge of the base is 6.3 feet, find the lateral area of the steeple.

21. A railroad tie is in the shape of a rectangular solid. It is 6 feet long, 8 inches wide, and 6 inches thick. The tie is to be dipped in creosote. What is the area of the treated surface?

22. How many square feet of copper are required to line the bottom and sides of a rectangular cistern, 12.7 feet long, 7.4 feet wide, and 4.8 feet deep?

23. Find the cost of building a stone wall 148 feet long, 5.4 feet high, and 1.5 feet thick at $4.50 a perch. (One perch = approximately 22 cubic feet.)

24—26. In Problems 13, 14, and 15, use the Pythagorean theorem to determine the slant height of each pyramid, and then find the total area of each pyramid.

27—28. In Problems 16 and 17, use the Pythagorean theorem to determine the altitude of each pyramid, and then find the volume of each pyramid.

8 | Cylinder

A cylinder is a very common geometric figure. Tanks and cans for various uses are frequently in the form of cylinders. If a rectangle is rotated about one side as an axis, the geometric solid formed is known as a right circular cylinder. The bases of the cylinder are circles. Each element of the curved surface is perpendicular to the bases. See Figure 19.

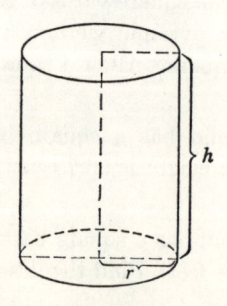

Figure 19

MEASUREMENT OF PLANE AND SOLID FIGURES 201

The volume of a circular cylinder equals π times the square of the radius times the altitude. Written as a formula, this statement is $v = \pi r^2 h$.

The lateral surface of a right circular cylinder equals the area of a rectangle whose altitude is the altitude of the cylinder and whose base is the circumference of the cylinder. This fact can be shown by joining the ends of a rectangular sheet of paper and observing Figure 20.

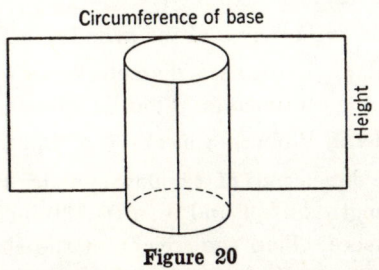

Figure 20

The lateral area or curved surface of a right circular cylinder equals the product of the altitude and the circumference of the base. The formula for this relationship is $A = 2\pi rh$.

Find the lateral area and volume of a right circular cylinder the radius of whose base is 6.3 inches and whose height is 7.8 inches.

Now:
$$A = 2(3.14)(6.3)(7.8)$$
$$A = (6.28)(6.3)(7.8)$$
$$A = (39.6)(7.8)$$
$$A = 308.88 \text{ or } 310 \text{ square inches}$$

Also:
$$v = \pi r^2 h$$
$$v = (3.14)(6.3)^2(7.8)$$
$$v = (3.14)(39.7)(7.8)$$
$$v = (125)(7.8)$$
$$v = 980 \text{ cubic inches}$$

EXERCISE 7

1. Find the volume of the cylinder the radius of whose base is 9.6 inches and whose height is 14.8 inches.

2. Find the lateral area of the cylinder described in Problem 1.

3. Write the formula for the total surface area of a right circular cylinder. (We mean here the lateral surface plus the top and bottom.)

4. Find the volume of a cylinder the radius of whose base is 12.4 feet and whose height is 15.6 feet.

5. Find the total surface area of the cylinder in Problem 4.

6. Find the volume of a cylinder the radius of whose base is 15.8 meters and whose height is 18.3 meters.

7. Find the total surface area of the cylinder in Problem 6.

8. A right circular cylinder has a diameter equal to 9.6 meters and a height equal to 425 centimeters. Find its volume.

9. If the cylinder in Problem 8 has no top, find its total area.

10. The outside dimensions of a hollow cylindrical cast-iron shaft are as follows: length 28 feet and diameter 30 inches. The inside diameter is 29 inches. Find the weight of the shaft if cast iron weighs 0.26 pound per cubic inch.

11. Find the capacity in tons of a circular silo 21 feet in diameter and 34 feet high, if 1 cubic foot of silage weighs 40.7 pounds.

12. Find the amount of lumber that will be needed for the lateral surface of a cylindrical silo 31 feet high and 16 feet in diameter.

*13. Hassocks are made in the form of right circular cylinders of diameter 28 inches and altitude 18 inches. What is the cost of construction if the leather covering costs $3.25 per square yard for the top and cylindrical surface, and $1.75 per square yard for the bottom? Other expenses will amount to 65 cents. Waste will amount to 12%.

14. How much paint will be needed for the lateral surface of a cylindrical silo whose diameter is 28 feet and whose height is 30 feet? Assume that 1 gallon will cover approximately 250 square feet.

15. What is the grinding surface of a grindstone that is $1\frac{1}{2}$ feet in diameter and 3 inches thick?

16. A cylindrical oil drum has a radius of 19 inches and a height of 42 inches. How many gallons does it hold?

*17. A lawn roller is made from a hollow iron cylinder whose inside diameter is 1.4 feet and whose length is 2.5 feet. If it weighs 38 pounds when empty what will it weigh when filled with water? Assume that water weighs 62.5 pounds per cubic foot.

9 | Cone

If a right triangle is rotated about one side as an axis, a right circular cone is formed. Thus in Figure 21, triangle ACB is rotated about AC, forming the cone. The lateral area of the surface of a cone is all of the area except that of the base. The lateral area equals one-half the product of the circumference of the base and the slant

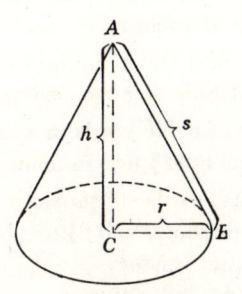

Figure 21

height. Thus $L = \pi rs$, in which r is the radius of the base, s the slant height, and L the lateral surface. Letting A be the total surface of the cone, we have $A = \pi rs + \pi r^2$ or $\pi r(s + r)$.

The volume of a cone equals one-third of the product of the area of the base of the cone and its altitude. The formula for volume is $v = \frac{1}{3}\pi r^2 h$, in which r again is the radius of the base and h is the altitude.

Find the total surface area of a cone the radius of whose base is 12.3 inches and whose slant height is 25.8 inches. Now:

$$A = \pi r(s + r) \quad \text{or} \quad A = (3.14)(12.3)(25.8 + 12.3)$$

Then $A = (38.62)(38.1)$ or $A = 1470.$ square inches.

Now let us find the volume of the above cone. But:

$$v = \frac{1}{3} \pi r^2 h, \quad \text{or} \quad v = \frac{1}{3} 3(.14)(12.3)^2(25.8)$$

or

$$v = \frac{1}{3}(3.14)(151.3)(25.8)$$

Now:

$$v = \frac{1}{3}(3.14)(3904) \quad \text{or} \quad v = \frac{1}{3}(12260.) = 4090 \text{ cubic inches}$$

EXERCISE 8

1. Find the volume of a cone, the radius of the base being 9 inches and the altitude 15 inches.

2. Find the volume of a cone, the radius of the base being 32.4 centimeters and the altitude 65.8 centimeters.

3. Find the number of cubic yards in a sandpile that is assumed to be a cone, if the cone is $12\frac{3}{4}$ feet in diameter and $3\frac{1}{2}$ feet high.

4. How many square yards of canvas are required to make a conical tent 9.6 feet in diameter and $8\frac{1}{4}$ feet high, allowing 3.5 square yards extra for seams and waste?

5. Find the formula for the radius of a cone when the volume and altitude are known.

6. Find the formula for the altitude of a cone when the volume and the radius of the base are given.

7. A right triangle with sides of 48.6 inches and 31.3 inches is revolved about the longer side. Find the lateral area of the cone generated.

8. In the above problem find the total area of the cone.

9. Find the volume of the cone described in Problem 7.

10. A right triangle with sides of 64.3 centimeters and 52.8 centimeters is revolved about the longer side. Find the lateral area of the cone generated.

11. In the above problem find the total area of the cone.

12. Find the volume of the cone described in Problem 10.

★13. A buoy is composed of a right circular cylinder of diameter 5.6 feet and altitude 9.8 feet surmounted by a cone whose altitude is 8.6 feet. Find the surface of the entire buoy.

14. The lateral area of a cone of revolution is 583.2π square centimeters, and the slant height is 48.6 centimeters. Find the altitude of the cone.

MEASUREMENT OF PLANE AND SOLID FIGURES   205

15. The lateral area of a cone of revolution is 257.6π square inches, and the slant height is 36.8 inches. Find the altitude of the cone.

16. The total area of an equilateral cone is 162π square meters. Find the radius of the base of the cone.

17. The total area of an equilateral cone is 288π square feet. Find the radius of the base of the cone.

18. The area of the base of a right circular cone is 31,400 square inches, and the altitude of the cone is 27.9 inches. Find the slant height.

19. A right triangle with legs of 21 feet and 28 feet is revolved about its longer leg as an axis. Find the lateral surface of the cone thus formed.

20. The radius of the base of a right circular cone is 11.6 feet, and the altitude is 14.7 feet. Find the total area of the cone.

21. Find the volume of the cone described in Problem 20.

10 | Sphere

The word sphere comes from the Greek *sphaira*, meaning a ball. A sphere is a solid included by a closed surface every point of

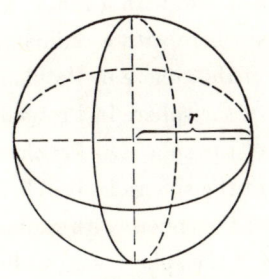

Figure 22

which is the same distance from a point within, called the center. Spheres are used very frequently in our highly mechanized age. As an example think of ball bearings in wheels.

A sphere is shown in Figure 22. The volume of a sphere is found by multiplying four-thirds times pi times the cube of the radius. Written as a formula this statement is $v = \frac{4}{3}\pi r^3$. The surface area of a sphere is found by multiplying four times pi times the square of the radius. This formula is $S = 4\pi r^2$.

Find the volume of a sphere the radius of whose base is 0.125 of an inch.

$$v = \tfrac{4}{3}\pi r^3$$
$$v = \tfrac{4}{3}(3.142)(0.125)^3$$
$$v = \tfrac{4}{3}(3.142)(0.0020)$$
$$v = \tfrac{4}{3}(0.0063)$$
$$v = 0.008 \text{ cubic inch}$$

Find the surface of the above sphere.

$$S = 4\pi r^2$$
$$S = 4(3.142)(0.125)^2$$
$$S = 4(3.142)(0.0156)$$
$$S = 4(.0490) = 0.196 \text{ of a square inch}$$

EXERCISE 9

1. Find the area of a sphere with a radius of 9 inches.
2. Find the area of a sphere with a radius of 7 meters.
3. Find the area of a sphere with a radius of 12 meters.
4. Find the area of a sphere with a radius of 10 inches.
5. Find the volume of the sphere in Problem 1.
6. Find the volume of the sphere in Problem 2.
7. Find the volume of the sphere in Problem 3.
8. Find the volume of the sphere in Problem 4.
9. Find the volume of the sphere with radius 14 feet.
10. Find the volume of the sphere with radius 16 meters.
11. Find the area of the sphere in Problem 9.
12. Find the area of the sphere in Problem 10.

*13. The moon is 2160 miles in diameter. Find its surface area in square miles.

14. The geography department in a college has twelve 10-inch globes. That is, the globes are 10 inches in diameter. Find the surface area of one of these globes.

MEASUREMENT OF PLANE AND SOLID FIGURES   207

15. How many cubic feet of gas will a spherical storage tank hold if its inside diameter is 48 feet?

16. How much leather is needed to cover a basketball 10 inches in diameter? Allow 15% for seams and waste.

17. If the radius of a sphere is doubled, what effect does this have on the volume?

18. A capitol dome is a hemisphere 80 feet in diameter. How many squares of paint will be needed for two coats of paint? (A square = 100 square feet.)

★19. A lead sphere 8 inches in diameter is melted and converted into lead shot $\frac{1}{8}$ inch in diameter. Approximately how many pellets of shot will there be?

20. If the radius of a sphere is multiplied by four, what effect does this have on the volume?

★21. How many spherical shot 0.18 inch in diameter can be made from a cylindrical bar of metal 4 inches in diameter and 3 feet long?

22. A malleable iron bar in the shape of a cylinder 4 feet long and 3 inches in diameter is drawn into iron wire $\frac{1}{16}$ inch in diameter. About how much more than a half mile of wire will there be?

23. How many balls $\frac{1}{8}$ inch in diameter can be made from a steel ball 9 inches in diameter?

24. Two spheres of radii 5 inches and 6 inches are melted and recast into a new sphere. Find the area of the new sphere.

EXERCISE 10
Review

1. Find the square root of 86.37.

★2. Find the length of the longest wire that can be stretched in a room that is 48 feet long, 38 feet wide, and 11 feet high.

3. A chemist has two acid solutions; one is 70% acid and the other is 20% acid. How much of each should he take in order to have 140 cubic centimeters of acid solution that is 60% acid?

4. Find the square root of 0.0676.

5. The Varsity Pharmacy has on hand a 10% and a 15% solution of argyrol. How should these be mixed in order to obtain 18 ounces of a 13% solution?

6. How many quarts of water should be added to one-half gallon of a full strength syrup in order to obtain a syrup that is 20% of full strength?

7. Solve for x: $3x + 15 = 45 - 7x$.

8. Find the interest on $845 at 7% from March 25, 1968 to October 24, 1968. Use ordinary interest for exact time,

9. Find the interest on $1536 at 8% from May 21, 1969 to December 16, 1969. Use ordinary interest for exact time.

10. Solve for x: $8x - 14 = 3x + 21$.

11. Divide 8.4567 by 93.8. Assume that both are approximate numbers.

12. Divide 384.69 by 2.78. Assume that both are approximate numbers.

13. Multiply 64.7 × 0.03586. Assume that both are approximate numbers.

14. Multiply 78.396 × 48.7. Assume that both are approximate numbers.

15. Add the approximate numbers 729.64 + 38.87 + 329 + 384.76 + 285.6.

16. Add the approximate numbers 896.73 + 38.75 + 652 + 748.69 + 438.8.

17. Subtract the approximate numbers 87.39 − 46.873.

18. Subtract the approximate numbers 658.07 − 47.68.

19. Al and Sam are carrying loads of 30 pounds and 40 pounds, respectively. How many pounds must be removed from Al's load and added to Sam's load so that Al will be carrying a load equal to $\frac{2}{3}$ of Sam's?

*20. In the center of a circular pool in ancient Peking there stood a tall straight reed that extended two arms' length above the level of the water. When a strong wind blew the reed, the tip touched the edge of the pool at the water level. If the radius of the pool was 8 arms' length, how deep was the pool?

21. $I = E/R$ is Ohm's Law in electricity. Solve this formula for R.

22. In Problem 21, if $I = 8$ amperes and $R = 30$ ohms, find E. The answer will be given in volts.

MEASUREMENT OF PLANE AND SOLID FIGURES ———————209

23. How many gallons of water will it take to fill an aquarium whose inside dimensions are 20 × 16 × 12 inches?

24. An automobile traveling at the speed of 65 miles per hour is traveling at how many feet per second?

25. If $1\frac{1}{2}$ inches on a map equals 60 miles, what distance will $5\frac{1}{4}$ inches represent?

26. How many 8 × 8 inch asphalt tiles should Mr. Appleby buy to cover the floor of his 20 × 25 foot recreation room, assuming that an additional one-tenth will take care of waste?

27. Find the proceeds from a loan of $1800 for a period of $1\frac{1}{2}$ years, if the charge is computed at $4\frac{1}{2}\%$ bank discount?

28. If $5\frac{1}{4}$ inches on a map equals 420 miles, how many inches would there be between the map representations of two points that are actually 640 miles apart?

29. Stone County Commissioners' Court passes a budget of $5,038,054 to be raised by taxing real estate assessed at $168,549,000. What is to be the tax rate per $100 valuation?

30. What is the annual simple interest rate on a loan of $750 if three years' interest amounted to $150.60?

31. If 17.6 pounds of iron, when oxidized, yields 27.8 pounds of iron oxide, how much iron oxide can be made from a ton of iron?

32. An electric dryer rated at 4.3 kilowatts takes 30 minutes to dry a load of wash. What is the cost per load if the local utility company charges 4 cents per kilowatt-hour?

33. If a gallon of paint covers 400 square feet, how many quart cans must be purchased to paint the ceiling and side walls of a room 19 × 25 feet and 8 feet high, if windows and doors take up 20% of the wall area?

34. If $C = \frac{5}{9}(F - 32)$, what is the Fahrenheit equivalent of 45°C?

35. Change $\frac{4}{500}$ to a decimal.

36. A suit, reduced 20% from its original price, is now on sale for $76.50. What was the original price of the suit?

37. A metal washer has an outer diameter of 9.0 centimeters, a hole of diameter 2.0 centimeters, and a thickness of 2 millimeters. How many cubic centimeters of metal does it contain?

38. Change $\frac{2}{5}\%$ to a decimal.

39. A 76 foot rope, when stretched from the top of a flagpole, touches the ground 28 feet from the base of the pole. How high is the pole?

40. How many cubic feet of oil will a spherical oil tank 54 feet in diameter hold?

41. Find the number of minutes in z hours.

42. Find the dimension of a square having the same area as a triangle whose base is 48 and altitude 24.

43. If Mac is y years old now, how old will he be 16 years from now?

44. Find the number of weeks in d days.

45. Solve $19.6 + 4x - 2.2 = 4.8 - 7$ for x.

46. Solve $\frac{3x}{7} = 63$ for x.

47. The Condemnation Board appraised a piece of land for the State Highway Department. If the land measures 4378 feet long and 126 feet wide, find how much the owner of the property will receive if the value is set at $1216 an acre.

48. If each of 35 long-playing records plays for 16 minutes 9 seconds, how long will it take to hear them all?

49. How much lumber is required for the lateral surface of a cylindrical silo 36 feet high and 19 feet in diameter?

*50. A spider is in the center of the floor of a building 40 yards long, 25 yards wide, and 12 feet high. How far will it walk by taking the shortest path to one of the upper corners of the ceiling?

51. Find the square root of 64.859.

*52. Hassocks are made in the form of right circular cylinders of diameter 26 inches and altitude 18 inches. Find the cost of construction if the leather covering costs $4.50 per square yard for the top and cylindrical surface, and $1.90 per square yard for the bottom. Waste should be figured at 12%. The filler and labor for each amounts to 92 cents.

53. Find the length of the diagonal of the rectangle whose length is 28 inches and whose width is 13.4 inches.

54. Find the square root of 964.73.

55. A rectangular oil can has a square base $14\frac{1}{2}$ inches by $14\frac{1}{2}$ inches and is 16 inches deep. How many gallons does it hold?

MEASUREMENT OF PLANE AND SOLID FIGURES   211

56. Find the length of the diagonal of the rectangle whose width is 54 meters and whose length is 73 meters.

57. Find the length of the diagonal of the rectangle whose length is 48 feet and whose width is 35 feet.

58. Subtract 9 gallons 3 quarts 1 pint from 14 gallons.

59. Subtract 4 yards 1 foot 9 inches from 7 yards 8 inches.

60. An almanac says that $46\frac{1}{2}$ quarts of milk weigh 100 pounds. How much will seven and three fourths gallons of milk weigh?

61. Write any number of four digits in descending order. Reverse the order of the digits and subtract. Reverse the order of the digits in the remainder and add to the remainder. The sum will always be the same regardless of the number selected at the beginning.

62. Write the natural number 112 as a number with the base 2.

63. Write the natural number 192 as a number with the base 2.

64. 11011011 is a number written in the binary system. Translate it into the decimal system.

65. 11110111101 is a number written in the binary system. Translate it into the decimal system.

Answers to Odd Numbered Problems

Chapter 2

Exercise 1, page 10

1. Thirty-eight   3. Eighty-four   5. Sixty-eight   7. Thirty-five
9. Nine hundred twenty-seven   11. Three hundred forty-seven
13. Five hundred ninety   15. Eight hundred six
17. Eight thousand, six hundred fifty-three
19. Seven thousand, three hundred sixty-four
21. Three thousand, six hundred thirty-five
23. Eight thousand, six hundred ninety-four
25. Two thousand, three   27. Eight thousand, ninety
29. Five thousand, six
31. Thirty-four thousand, six hundred seventeen
33. Nineteen thousand, six hundred fifty-seven
35. Twenty-nine thousand, seven hundred sixty
37. Eighty-seven thousand, four hundred twenty-three
39. Five hundred thirty-six thousand, eight hundred fifty-three
41. Three million, five hundred eighty-nine thousand, four hundred sixty-two
43. Forty-five million, three hundred ninety-two thousand seven hundred sixty-two

45. Seven hundred forty-five million, six hundred thirty-two thousand, four hundred twenty-nine

47. Sixty-eight trillion, four hundred thirty-five billion, six hundred twenty-nine million, thirty-six thousand, two hundred fifty-nine

49. Four hundred thirty-seven trillion, eight hundred twelve billion, nine hundred forty-seven million, six hundred thirty-eight thousand, two hundred seventeen

51. The 6 is in the hundreds position and represents 600

53. The 6 is in the thousands position and represents 6000

55. The 4 is in the ten thousands position and represents 40,000; the 3 is in the thousands position and represents 3000; thus the 43 represents 43,000

57. The 9 is in the thousands position and represents 9000; the 6 in the hundreds position and represents 600; thus the 96 represents 9600

59. The 5 is in the ten thousands position and represents 50,000; the 4 is in the thousands position and represents 4000; thus the 54 represents 54,00

61. 642   **63.** 835,872   **65.** 38,639,008,065,642
67. 962,431,087,009   **69.** 4,004,004,004,400

Exercise 2, page 15

1. 4; 7; 15; 10; 4; 10; 12; 17; 9; 15
3. 387; 416; 401; 493; 492; 526
5. 10,041; 9962; 17,602; 8,817; 13,023
7. 35; 157; 2923; 2345; 33,572
9. 265,812   **11.** $2767.04; $29,167.40
13. 45,972,406   **15.** 3,374,696,956
17. 21; 27; 78; 87; 79   **19.** 252; 607; 175; 221; 324
21. 5051; 1636; 1842; 3187; 2781
23. 7547   **25.** 241,969   **27.** 4,115,669   **29.** 81,198
31. 266,309,090,119   **33.** 240; 48   **35.** 1687
37. 2 hours   **39.** $156.96   **41.** 252 pounds
43. 32,291 pounds

Exercise 3, page 21

1. 21; 9; 24; 40; 54; 10; 56; 0; 25; 0
3. 92; 126; 46; 132; 236   **5.** 360; 962; 1881; 3612; 2925

ANSWERS TO ODD NUMBERED PROBLEMS — 215

7. 20,501　　**9.** 63,516　　**11.** 684,411　　**13.** 3,908,862
15. 4,721,704　　**17.** 70,609,604　　**19.** 36,013,572　　**21.** 703,627,986
23. Five hundred forty-seven; eight thousand, six hundred thirty-two
25. 59,365,466,736,423
27. (a) $6.48; (b) $7.50; (c) $15.64
29. $7.58　　**31.** 2,646 cubic feet　　**33.** 77,760 pounds
35. 42,240 feet　　**37.** 1680 square feet　　**39.** 6048 buttons

Exercise 4, page 24

1. 5; 4; 6; 7; 8　　　　**3.** 49; 37; 39; 47; 26
5. 7 r 1　　**7.** 21 r 24　　**9.** 16 r 7　　**11.** 61
13. 37　　**15.** 55　　**17.** 24　　**19.** 59
21. 96 r 23　　**23.** $46\frac{58}{63}$　　**25.** 256 r 179　　**27.** $74\frac{554}{3025}$
29. $222\frac{381}{2108}$　　**31.** $19\frac{5869}{19321}$　　**33.** $3\frac{1,339,858}{1,956,141}$　　**35.** 14
37. 19 miles per gallon　　**39.** $426
41. 24 inches　　**43.** 990

Exercise 5, page 27

1. VIII　　**3.** XXV　　**5.** XLII　　**7.** XCIV
9. LXXXVII　　**11.** CXXVI　　**13.** DCCCLIX
15. $\overline{\text{IX}}$DCCLXVIII　　**17.** $\overline{\text{CCCLXXXVI}}$ DCCCXLII
19. 25　　**21.** 32　　**23.** 47　　**25.** 44
27. 99　　**29.** 327　　**31.** 458　　**33.** 7458
35. 456,744

Exercise 6, page 28

1. Seven thousand, six hundred forty-nine

3. Sixty-five million, three hundred ninety-two thousand, four hundred sixty-seven

5. Forty-seven billion, six hundred twenty-nine million, five hundred ninety-four thousand, eight hundred twenty-six

7. Seven hundred thirty-six million, five hundred forty-three thousand, six hundred twenty-eight dollars

9. 301 eggs　　**11.** 18,968,413,327,707
13. 24,429　　**15.** 3,828,639,306
17. Monday, $1452; Tuesday, $1145; Wednesday, $1968; Thursday,

$1998; Friday, $1706; Saturday, $1428. Evers, $2051; Freeman, $1475; Howell, $1977; Jarman, $2595; Keown, $1599. Total sales for week $9,697

19. $211.31

21. Eight hundred sixty-five billion, four hundred ninety-two million, five hundred thirty-nine thousand, nine hundred twenty-seven

23. $7550.62 **25.** 15 inches

Chapter 3

Exercise 1, page 35

1. $\frac{6}{8}; \frac{9}{12}; \frac{75}{100}; \frac{24}{32}$ **3.** $\frac{4}{12}; \frac{9}{12}$

5. $\frac{10}{15}; \frac{9}{15}$ **7.** $\frac{9}{12}; \frac{10}{12}$ **9.** $\frac{21}{35}; \frac{15}{35}$

11. $\frac{1}{5}, \frac{4}{5}, \frac{2}{5}, \frac{5}{5}, \frac{8}{5}, \frac{9}{5}, \frac{2}{5}$ **13.** $\frac{6}{9}, \frac{5}{9}, \frac{4}{9}, \frac{1}{9}, \frac{21}{9}, \frac{9}{9}, \frac{27}{9}$

15. $\frac{8}{32}, \frac{10}{32}, \frac{12}{32}, \frac{44}{32}, \frac{16}{32}, \frac{24}{32}, \frac{5}{32}, \frac{96}{32}$

17. 6 pints; 40 pints   **19.** $\frac{3}{4}$ of a foot; $\frac{1}{12}$ of a foot

21. $\frac{45}{144}; \frac{64}{144}$ **23.** 13 quarts   **25.** 92 ounces

27. $\frac{25}{30}; \frac{40}{48}; \frac{30}{100}; \frac{108}{300}$ **29.** $\frac{48}{80}; \frac{25}{80}$

Exercise 2, page 39

1. $\frac{2}{3}$ **3.** $\frac{1}{3}$ **5.** 1   **7.** $\frac{1}{3}$

9. $\frac{2}{5}$ **11.** $\frac{5}{4}$ or $1\frac{1}{4}$ **13.** $\frac{51}{56}$ **15.** $\frac{1}{4}$

17. $\frac{5}{16}$ **19.** $\frac{53}{28}$ or $1\frac{25}{28}$ **21.** $\frac{379}{240}$ or $1\frac{139}{240}$

23. $\frac{37}{80}$ **25.** $\frac{279}{144}$ or $1\frac{135}{144}$ **27.** $\frac{23}{420}$

29. $\frac{251}{90}$ or $2\frac{71}{90}$ **31.** $\frac{9,241}{11,628}$ **33.** $\frac{5}{72}$

ANSWERS TO ODD NUMBERED PROBLEMS — 217

35. Ralph, 9; Reb, 6; Ruel, 2   **37.** $\dfrac{1}{8}$

39. $\dfrac{1}{2}$ **41.** $\dfrac{4}{7}$; $\dfrac{9}{16}$; $\dfrac{5}{18}$; $\dfrac{3}{25}$

Exercise 3, page 43

1. $\dfrac{2}{5}$ **3.** $\dfrac{2}{7}$ **5.** $\dfrac{1}{3}$ **7.** $\dfrac{3}{10}$

9. $\dfrac{1}{8}$ **11.** $\dfrac{1}{14}$ **13.** $\dfrac{2}{9}$ **15.** $\dfrac{2}{105}$

17. $\dfrac{3}{238}$ **19.** $\dfrac{1}{18}$ **21.** $\dfrac{352}{2555}$ **23.** $\dfrac{5}{48}$

25. $\dfrac{1}{16}$ **27.** $\dfrac{9}{80}$ **29.** $\dfrac{20}{1749}$ **31.** $\dfrac{23}{630}$

33. $\dfrac{1}{32}$ **35.** $\dfrac{13}{405}$ **37.** $\dfrac{435}{42{,}172}$ **39.** $3515.62

41. $72\tfrac{5}{6}$ feet   **43.** 504 ounces   **45.** $6571.25

47. $179\tfrac{5}{32}$ cubic inches   **49.** $1\tfrac{2}{55}$

51. 14 feet by 20 feet   **53.** 40

Exercise 4, page 47

1. $\dfrac{1}{2}$ **3.** $\dfrac{2}{3}$ **5.** $\dfrac{5}{6}$ **7.** $\dfrac{1}{9}$

9. $\dfrac{1}{20}$ **11.** $\dfrac{1}{20}$ **13.** $\dfrac{5}{4}$ **15.** $\dfrac{2}{5}$

17. 1   **19.** $\dfrac{7}{18}$ **21.** $\dfrac{2}{5}$ **23.** $1\tfrac{25}{18}$

25. 1   **27.** $\dfrac{3}{20}$ **29.** $\dfrac{3}{8}$ **31.** $\dfrac{2}{5}$

33. $\dfrac{4}{9}$ **35.** 1   **37.** $1\tfrac{39}{65}$ **39.** $\dfrac{19}{21}$

41. $\dfrac{16{,}720}{21{,}141}$ **43.** $5\tfrac{13}{55}$ **45.** 25 pieces

47. $17\tfrac{2}{3}$ miles per gallon   **49.** $1\tfrac{2}{3}$ hours

51. 3   **53.** 3 hours

Exercise 5, page 51

1. $11\frac{1}{4}$
3. $29\frac{11}{16}$
5. $43\frac{4}{15}$
7. $38\frac{13}{30}$
9. $26\frac{3}{4}$
11. $36\frac{23}{30}$
13. $599\frac{9}{16}$
15. $979\frac{1}{24}$
17. $354\frac{39}{40}$
19. $2256\frac{88}{105}$
21. $4\frac{1}{8}$
23. $9\frac{57}{80}$
25. $39\frac{293}{476}$
27. $29\frac{19}{68}$
29. $3\frac{7}{20}$
31. $15\frac{416}{495}$
33. $8\frac{137}{255}$
35. $3\frac{5}{84}$
37. $12\frac{59}{60}$
39. $28\frac{17}{168}$
41. $\frac{2}{15}$
43. $\frac{151}{147}$
45. $4\frac{430}{1428}$
47. $5\frac{503}{630}$
49. $7\frac{1}{12}$ feet
51. $6\frac{1}{4}$ inches
53. $26\frac{1}{4}$ yards
55. $21\frac{2}{3}$ hours
57. $61\frac{1}{3}$ feet

Exercise 6, page 55

1. 3
3. $4\frac{19}{20}$
5. 9
7. $4\frac{2}{3}$
9. 35
11. 4
13. $15\frac{3}{4}$
15. $18\frac{1}{8}$
17. $14\frac{2}{5}$
19. 8
21. $48\frac{7}{8}$
23. $118\frac{1}{8}$
25. $329\frac{11}{14}$
27. $1013\frac{2}{5}$
29. $6\frac{6}{11}$
31. $4\frac{4}{5}$
33. $8167\frac{200,359}{310,023}$
35. 1925 miles
37. 1359.375 pounds
39. $11\frac{2}{3}$ yards; $2\frac{5}{6}$ yards
41. $24\frac{3}{4}$ inches
43. 538.3125 bushels
45. 20 cups flour; 18 cups baking powder; 4 cups shortening; 10 cups sugar

Exercise 7, page 58

1. $\frac{4}{3}, 2, \frac{1}{3}, \frac{2}{3}, \frac{3}{4}$
3. $\frac{3}{2}$
5. $\frac{7}{3}$
7. $\frac{7}{4}$
9. $\frac{5}{3}$
11. $\frac{27}{35}$
13. $7\frac{91}{243}$
15. $\frac{5}{4}$
17. $\frac{7}{3}$
19. $\frac{7}{2}$
21. $\frac{7}{3}$
23. $1\frac{53}{118}$
25. $3\frac{547}{675}$
27. $2\frac{41,225}{88,668}$
29. $14\frac{698}{2421}$
31. $\frac{13}{32}$
33. $\frac{1}{5}$
35. $\frac{56}{627}$
37. $19\frac{1171}{2033}$
39. $\frac{15}{64}$
41. $\frac{9}{4}$
43. $\frac{10}{7}$
45. $59\frac{31}{35}$ miles per hour
47. $1\frac{2}{27}$ feet
49. $18\frac{2}{3}$ feet by $11\frac{1}{6}$ feet
51. 32
53. $5\frac{10}{33}$ feet

ANSWERS TO ODD NUMBERED PROBLEMS — 219

Exercise 8, page 60

1. $2\frac{1}{12}$
3. $1\frac{11}{24}$
5. $\frac{4}{3}$
7. $2\frac{129}{140}$
9. $\frac{1}{4}$
11. $\frac{3}{8}$
13. $\frac{19}{40}$
15. $\frac{9}{80}$
17. $1\frac{1}{7}$
19. $1\frac{17}{18}$
21. $11\frac{3}{4}$
23. $51\frac{1}{12}$
25. $67\frac{1}{5}$
27. 180
29. 77
31. $2582\frac{17}{24}$
33. $1\frac{87}{88}$
35. $3\frac{1}{2}$
37. $5\frac{113}{183}$
39. 125 stamps
41. 480 pounds
43. $4.80
45. 14 feet by 15 feet
47. $4\frac{3}{4}$ hours
49. Place two men in a group and ask one of them to hold a dollar bill. Give the third man a dollar, the fourth man a dollar, the fifth man a dollar, the sixth man a dollar, and the seventh man a dollar. There is now one dollar remaining. Go back to the first two men, and give the remaining dollar to the one who is not holding a dollar bill.
51. 360
53. $38\frac{2}{21}$ pounds
55. $\frac{2}{7}$
57. $47

Chapter 4

Exercise 1, page 69

1. 0.8
3. 0.08
5. 2.42
7. 11.5
9. 15.24
11. 17.629
13. 1562.94
15. 156.46
17. 0.8
19. 14.3
21. 1.122
23. 21.57
25. 0.895
27. 8.07
29. 14.196
31. 0.13
33. 0.05
35. 0.27
37. 0.463
39. 13.2
41. 44.209
43. 0.12259
45. 0.175
47. 2.88
49. 35,144.8052
51. 212,511.29947
53. 27,118.6473
55. 9,178.877
57. Four hundred eighty-six and ninety-three hundredths
59. Nine thousand, seven hundred sixty-three and seven hundred eighty-nine thousandths
61. Five hundred ninety-two thousand, six hundred forty-eight and seven thousand nine hundred sixty-three ten thousandths
63. Seven hundred nine thousand, five hundred forty-seven and six thousand three hundred two ten thousandths

65. Fifty-nine million, seven hundred fifty-six thousand, nine hundred eighty-seven and sixty-seven thousand eight hundred thirty-four hundred thousandths

67. 3846.73 **69.** 45,647.829 **71.** 394,469.6907
73. 545,675,537.67467   **75.** 16,648.9847
77. 621,390.64722 **79.** $132.46 **81.** $7499.74
83. 72.734 **85.** 768.407 **87.** 30.027
89. 0.8, 0.9, 0.875, 0.5625, 0.2 **91.** 1.5, 1.125, 4.75, 5.2, 3.125
93. $\dfrac{9}{25}, \dfrac{17}{40}, \dfrac{5}{8}, \dfrac{24}{25}, \dfrac{1}{16}$ **95.** $\dfrac{6}{5}, \dfrac{907}{20}, \dfrac{133}{8}, \dfrac{171}{8}, \dfrac{561}{4}$

Exercise 2, page 74

1. 3.6 **3.** 43.803 **5.** 36.98 **7.** 0.24
9. 0.035 **11.** 0.24 **13.** 0.648 **15.** 0.5561
17. 0.7191 **19.** 0.02214 **21.** 0.002961 **23.** 1,144.04
25. 4339.831 **27.** 5791.3262 **29.** 336,908.874
31. 537,972.792 **33.** 602,971.866 **35.** $3.68
37. $23.94 **39.** 41.5 miles **41.** $42.75 **43.** $1.53
45. $1,221,022,160 **47.** $1,809,554,208
49. $9.04 **51.** $318.62 **53.** $103,477.50

Exercise 3, page 79

1. 0.1 **3.** 1.7 **5.** 7.3 **7.** 0.7
9. 5.3 **11.** 0.43 **13.** 7.58 **15.** 16.07
17. 0.41 **19.** 8.68 **21.** 0.645 **23.** 7.050
25. 37.783 **27.** 0.588 **29.** 3.684 **31.** 87.49
33. 136.78 **35.** 1642.68 **37.** 3499.74 **39.** 36.76
41. 607.785 **43.** 326.846 **45.** 2364.956 **47.** 927.262
49. 64.297 **51.** 3618.594 **53.** 5783.676 **55.** 2.23
57. 0.95 **59.** 0.02 **61.** 0.12 **63.** 0.07
65. 34.16 **67.** $16.43
69. 68 **71.** 529 statute miles
73. 135.78 pounds copper, 24.18 pounds tin, 26.04 pounds zinc
75. 36.52 knots
77. Place 3 figs in one cup and 7 in another, and the place one of these in the third cup.
79. $17,478.94

ANSWERS TO ODD NUMBERED PROBLEMS — 221

Chapter 5

Exercise 1, page 86

A

1. 5 ft. 5 in.
3. 5 bu. 1 pk.
5. 6 hr. 10 min.
7. 11 qt. 2 pt. 5 oz.
9. 13 hr. 23 min. 50 sec.
11. 12
13. 16
15. 2
17. 5
19. 11 ft. 10 in.
21. 16 yd. 2 ft. 4 in.
23. 21 lb. 2 oz.
25. 16 hr. 46 min.
27. 17 yd. 1 ft. 6 in.
29. 54 mi. 2880 ft.

B

1. 22 gal. 2 qt.
3. 38 gal.
5. 22 lb. 13 oz.
7. 21 yr. 3 mo.
9. 31 yr. 6 mo.
11. 27 wk. 4 hr.
13. 3 yd. 1 ft. 2 in.
15. 2 qt. 2 pt.
17. 3 bu. 5 pk.
19. 2 gal. 3 qt.
21. 199 in.
23. 78 pt.
25. 149 oz.
27. 528 qt.
29. 24 cups

C

1. 217,800 square feet
3. 879 square yards
5. 22 pounds 7 ounces
7. 36 feet 6 inches
9. 17 gallons
11. 182 quarts
13. 2088 cubic yards
15. 106 rods and 15 feet
17. 7 bushels 2 pecks 2 quarts
19. 3 tons 754 pounds
21. 313 quarts
23. 399 pounds 5 ounces
25. 211 gallons 2 quarts
27. 223 hours 47 minutes 54 seconds
29. 42 pounds 8 ounces

D

1. 8 feet 9 inches
3. 3 hours 23 minutes
5. 9 gallons 1 pint
7. 79 feet 4 inches
9. 332 pounds 7 ounces
11. 457 hours 13 minutes
13. 1216 square feet 39 square inches
15. 273,259 square feet 45 square inches
17. 32 pounds 3 ounces
19. 20 hours 6 minutes 30 seconds
21. $17\frac{1}{2}$
23. 7 hours 51 minutes
25. 27 pounds
27. 19 feet 1 inch
29. $128.27

E

1. 6 pounds 8 ounces
3. 4800
5. 28 feet
7. 29 revolutions in 1 second
9. $93.04
11. 165 gallons
13. 3360 revolutions per minute
15. 134,400
17. 540 revolutions

Exercise 2, page 95

1. 100 centimeters
3. 10 millimeters
5. 40 centimeters
7. 1000 grams
9. 4 grams
11. 100 centigrams = 1 gram
13. 100,000 centigrams
15. 20,000 centimeters
17. 1000 milliliters
19. 100,000 centiliters
21. 400 milliliters
23. 83 centimeters 2 millimeters
25. 11 meters 12 decimeters 3 centimeters
27. 0.8 kilogram
29. 100 centigrams
31. 45,000 milligrams
33. 0.4 kilogram
35. 540,000 centimeters
37. 1 centimeter
39. Move decimal point three places to the right or multiply by 1000.
41. Move decimal point three places to the left or divide by 1000.
43. 91 kilometers
45. 800 centimeters
47. 40
49. 0.09 meter
51. 100,000,000 centigrams

Exercise 3, page 101

1. 80,000
1,750,000
380,000,000
0.00000000197
740,000,000,000
0.00009
0.00046
1,930,000,000,000
57,000,000,000
0.000000000000000000000164
3. 29,980,000,000
0.0000000000000000000000006547
0.000000000477
109,737
0.0000000000000000000000001650

ANSWERS TO ODD NUMBERED PROBLEMS   **223**

Exercise 4, page 104

1. 101   **3.** 11001   **5.** 110001   **7.** 111110
9. 1010110   **11.** 1101011   **13.** 1011110111
15. 3   **17.** 14   **19.** 16   **21.** 9
23. 31   **25.** 30   **27.** 1721

Exercise 5, page 106

1. 157.5 inches   **3.** 472.4 inches   **5.** 285.4 inches
7. 3.346 inches   **9.** 2.362 inches   **11.** 22.97 feet
13. 984.2 feet   **15.** 244.8 feet   **17.** 26.10 miles
19. 226.8 miles   **21.** 2.03 meters   **23.** 1.66 meters
25. 2.66 meters   **27.** 52.92 pounds   **29.** 18.52 pounds
31. 4.54 kilograms   **33.** 3.75 kilograms   **35.** 90.8 kilograms

B

1. 393.7 inches   **3.** 8040 meters   **5.** 402 meters
7. 180.4 feet   **9.** 33,300 gallons per hour
11. 0.3977 mile   **13.** 114 grams per second
15. 2440 millimeters   **17.** 91.4 meters   **19.** 28.3 grams
21. 0.914 meter   **23.** 14.6 meters   **25.** 30,600 meters
27. 155.8 feet   **29.** 28.02 feet   **31.** 0.511 kilogram
33. 4.26 kilogram   **35.** 14.77 pounds   **37.** 390.7 cubic inches
39. 173 centimeters   **41.** 0.8114 pounds   **43.** 5.08 centimeters
45. 982 kilometers   **47.** 22.7 kilograms

Exercise 6, page 108

1. 157.59   **3.** 504 r 66
5. Nine billion, six hundred thirty-four million, eight hundred forty-six thousand, five hundred thirty-eight dollars

7. 6.79×10^8 **9.** 6,000,000,000   **11.** 0.482
13. 10010010   **15.** 237   **17.** $8 + \frac{3}{4}$ **19.** 17,400
21. 152   **23.** 1084.86   **25.** 770 square feet
27. 2.68   **29.** 21 inches   **31.** 54.8 pounds
33. 89 kilograms   **35.** $\frac{27}{82}$ **37.** $\frac{4}{9}$

39. 112 **41.** 5760 **43.** $100
45. MCMLXX **47.** 1948 **49.** 6.64×10^5
51. 3,240,000,000,000,000,000,000,000
53. 4383 **55.** 1010110001

Chapter 6

Exercise 1, page 114

1. 0.03   **3.** 0.05   **5.** 0.045   **7.** 0.62
9. 0.74   **11.** 1.45   **13.** 6.40   **15.** 32.60
17. 28.73   **19.** 0.3973   **21.** 6.375   **23.** 1.537
25. $40.39\frac{5}{8}$ **27.** 34.6975   **29.** 0.17   **31.** 0.075

Exercise 2, page 115

1. 3%   **3.** 6.25%   **5.** 8.33%   **7.** $6\frac{1}{4}$%
9. $7\frac{3}{4}$%   **11.** 68.73%   **13.** 37.43%   **15.** 765.4%
17. 8693%   **19.** $86,437\frac{1}{4}$%   **21.** 6.857%   **23.** $4932\frac{1}{2}$%
25. 69,237%   **27.** 75%; 25%   **29.** 25.2%

Exercise 3, page 116

1. $\dfrac{3}{50}$ **3.** $\dfrac{1}{25}$ **5.** $\dfrac{2}{5}$ **7.** $\dfrac{7}{4}$

9. $\dfrac{11}{200}$ **11.** $\dfrac{17}{400}$ **13.** $\dfrac{93}{200}$ **15.** $\dfrac{143}{400}$

17. $\dfrac{249}{200}$ **19.** $\dfrac{601}{400}$ **21.** $\dfrac{1269}{200,000}$ **23.** $\dfrac{11}{250}$

25. $\dfrac{1}{8}$; 18

Exercise 4, page 117

1. 40%   **3.** 37.5%   **5.** 375%   **7.** 800%
9. 187.5%   **11.** 125%   **13.** 600%   **15.** 40,000%
17. 1775%   **19.** 50%   **23.** 0.25%, $\dfrac{1}{75}$, $\dfrac{1}{10}$, $\dfrac{19}{24}$

25. $\dfrac{2}{5}$, $0.4\dfrac{3}{5}$, 46.5%, $\dfrac{17}{16}$ **27.** $\dfrac{1}{6}$; 16.67%

ANSWERS TO ODD NUMBERED PROBLEMS

Exercise 5, page 120

1. 11.9	**3.** 27.95	**5.** 140.01	**7.** 5.32%
9. 36.14%	**11.** 6.62%	**13.** 188.10	**15.** 158.06
17. 83.70	**19.** 1349.58	**21.** 3117.80	**23.** 38.30%
25. 165.04%	**27.** 246.15%	**29.** 573.38%	**31.** 960.96
33. 1756.02	**35.** 3029.14	**37.** 123.21	**39.** 138.75
41. 150.96	**43.** $2762.50	**45.** $7128	**47.** $12,500
49. $6.37	**51.** 954 pounds		

53. 20% cement; 40% sand; 40% gravel
55. 112,500 pounds
57. 2.02% chromium; 2.86% lead; 11% rubber
59. 2.1 pounds **61.** 3.59% **63.** 426.67 tons **65.** 14.61 pounds

Exercise 6, page 126

1. 23.1%	**3.** $14.57	**5.** $14.10	**7.** $42.86
9. 7.55%	**11.** $132.41	**13.** $5.43	**15.** $3829
17. $8.95		**19.** $2.90; 16.7%; 83.3%	
21. $91.20	**23.** $6.50	**25.** $30.37	**27.** $24.75 error
29. $11,600	**31.** $499.91		
33. It makes no difference		**35.** $356.51	

Chapter 7

Exercise 1, page 136

1. One pound
3. Ten quarts
5. One hundred feet
7. One foot
9. One tenth of a mile
11. One dollar
13. One dollar
15. One hundredth of an inch
17. One thousandth of a gallon
19. One fourth of a foot
21. Four
23. Five, four, or three depending on the unit of measure
25. Six 27. Six 29. One 31. One
33. Seven 35. Three 37. Seven 39. Six

B

1. 88	**3.** 76	**5.** 7.6	**7.** 535
9. 458	**11.** 985	**13.** 6793	**15.** 486.7

17. 73.05	**19.** 2585.6	**21.** 36,238	**23.** 89.348
25. 8.6	**27.** 5.8	**29.** 28.6	**31.** 14.9
33. 32.7	**35.** 296.8	**37.** 7969.9	**39.** 46.8
41. 8.57	**43.** 9.48	**45.** 12.26	**47.** 53.37
49. 128.97	**51.** 4532.54		

Exercise 2, page 138

1. 28.6   **3.** 549.9

5. Twenty-eight and sixty-five hundredths; five and four tenths; seven hundred thirty-eight thousandths; seven and five hundred forty-three thousandths; four hundred sixty-nine and seventy-three hundredths; thirty-seven and eight tenths; five hundred forty-nine and nine tenths

7. 591.24   **9.** 6928.858   **11.** 68,542.4   **13.** 117.8 miles
15. 22,772.3 miles
17. 48.0 inches; 0.4 inch below average
19. 2141.1 feet

Exercise 3, page 144

1. 14   **3.** 410   **5.** 360   **7.** 100
9. 5440   **11.** 4800   **13.** 32,100   **15.** 8468.7
17. 42,642   **19.** 24,000

21. Five thousand, six hundred ninety-three and seventy-eight hundredths; four and two tenths; twenty-four thousand

23. 16,000   **25.** 718,000   **27.** 34.6   **29.** 14,600
31. 653,000   **33.** 9.0   **35.** 1.8   **37.** 3.74
39. 26.41   **41.** 6.424   **43.** 20.943   **45.** 252.3
47. 8.3924   **49.** 4.725301   **51.** 314.58

53. Twenty-four thousand, nine hundred seventy-six; seven million, eight hundred fifty-six thousand, nine hundred eighty-two; three hundred fourteen and fifty-eight hundredths

55. 15.8048

57. Seventeen and five hundred forty-nine thousandths; three hundred fifty-nine and seventy-eight hundredths; twenty and five hundred one thousandths

59. 11.5 acres   **61.** 19.1 square feet   **63.** $414\tfrac{3}{8}$ square feet
65. 66,940 acres   **67.** 27.9 inches   **69.** 1800 pounds
71. 75%   **73.** 49 yards   **75.** $9.45   **77.** $1063.67

ANSWERS TO ODD NUMBERED PROBLEMS — 227

Chapter 8

Exercise 1, page 151

1. Five x square; 5; 2
3. Sixteen x square z cube; 16; 2; 3
5. Sixty-two a square b cube c to the fourth; 62; 2; 3; 4
7. Three fifths x cube y square; $\frac{3}{5}$; 3; 2
9. Seventeen a cube b to the fourth c to the fifth; 17; 3; 4; 5
11. $5x^3y^2$ 13. x^6 15. $64a^2b^3$ 17. $196x^8y^3$
19. $108a^5b^5$ 21. 8   23. 9   25. 9
27. 2   29. 35   31. 6   33. 1664
35. 28   37. 120   39. $\frac{8}{7}$

Exercise 2, page 154

1. -21 3. -33 5. 15   7. 22
9. -4 11. -6 13. $101\frac{1}{8}$ 15. $-288\frac{11}{40}$
17. 23   19. 23   21. -65 23. -43
25. $62\frac{1}{8}$ 27. -456.6 29. 32   31. 40
33. 102   35. -175 37. -336 39. -63.366
41. $24{,}035\frac{17}{32}$ 43. 4   45. -9 47. 6
49. -5 51. 0.02

Exercise 3, page 160

1. 3   3. 6   5. -5 7. 9
9. 4   11. 5   13. $12\frac{1}{2}$ 15. $27\frac{1}{4}$
17. 40   19. $45\frac{5}{7}$ 21. $8\frac{3}{4}$ 23. $5\frac{3}{20}$
25. $47\frac{1}{4}$ 27. $29\frac{1}{3}$ 29. 4.1   31. 3.08
33. 7   35. $-2\frac{2}{3}$ 37. $13\frac{1}{3}$ 39. $5\frac{3}{5}$
41. $29\frac{11}{21}$ 43. $-\frac{11}{15}$ 45. $2\frac{67}{121}$ 47. 21
51. 4   53. 104
55. $8000 Branch; $32,000 Patton
57. 12   59. $3, each girl; $9, each boy
61. $25.80

Exercise 4, page 163

1. 6	**3.** 4	**5.** 12.56	**7.** 88
9. 28	**11.** 411	**13.** 25	**15.** 78.5
17. 452.16	**19.** 918	**21.** 144	**23.** 2898
25. 382	**27.** 64	**29.** 6	**31.** 5
33. 468	**35.** 15	**37.** 9	

Exercise 5, page 168

1. 99 days	**3.** 114 days	**5.** 294 days	**7.** 95 days
9. 79 days	**11.** 238 days	**13.** 116 days	**15.** 173 days
17. 187 days	**19.** 236 days	**21.** 179 days	**23.** 227 days
25. 181 days	**27.** 246 days	**29.** 195 days	**31.** 232 days
33. $47.91	**35.** $12.75	**37.** $164.33	**39.** $540
41. 166 days	**43.** 6.98%	**45.** $2137.72	
47. $1755; $1800		**49.** $4417.50; $4500.00	
51. $437.66			

Exercise 6, page 172

1. $\dfrac{1}{4}$	**3.** $\dfrac{2}{3}$	**5.** $\dfrac{2}{5}$	**7.** 16
9. $\dfrac{22}{33}$	**11.** $\dfrac{887}{721}$	**13.** $\dfrac{3}{32}$	**15.** $\dfrac{1}{9}$
17. $\dfrac{1}{180}$	**19.** $\dfrac{17}{9000}$	**21.** $\dfrac{5}{6}$	**23.** $\dfrac{1}{2}$
25. $\dfrac{8}{7}$	**27.** $\dfrac{5}{3}$	**29.** $\dfrac{1}{1,267,200}$	

Exercise 7, page 176

1. 36 **3.** $\dfrac{5}{3}$ **5.** 18 **7.** 23
9. −5 **11.** 40 **13.** $6\tfrac{2}{3}$ **15.** 14
17. $2\tfrac{1}{4}$ **19.** 12 kilograms **21.** 70.2 miles per hour
23. 1.4 quarts per second **25.** 2.8 pounds
27. 171.4 pounds **29.** 210 pounds
31. The volume is multiplied by 8
33. 30; 10

Chapter 9

Exercise 1, page 181

1. 69.4 yards
3. 73.6 rods
5. $p = 2l + 2w$
7. 242.8 feet
9. 778.5 feet
11. 16.2
13. πr
15. $27\frac{9}{11}$ feet by $6\frac{2}{11}$ feet
17. 26 by 6
19. 154 inches by 110 inches

Exercise 2, page 184

1. 51.9 square inches
3. 700 square inches
5. 622 square feet
7. 61.14 acres
9. 11,000,000 bricks
11. 6 square yards
13. $260
15. 3 to 2
17. 102 feet
19. 240 inches

Exercise 3, page 186

1. 68
3. 27,000 square feet
5. 51 square centimeters
7. 9.71 square kilometers
9. 15,000 square inches
13. 1600 square inches
15. 2.45 acres

Exercise 4, page 192

1. 8.0
3. 12.0
5. 51.00
7. 85.00
9. 106.00
11. 117.00
13. 134.00
15. 155.00
17. 35.9000
19. 95.6000
21. 102.5000
23. 946.8254
25. 0.085
27. 0.1668
29. 21.0
31. 24 feet
33. 6 feet
35. 8 inches
37. 18 feet

Exercise 5, page 194

1. 200 square inches
3. 13 square inches
5. 2530 square meters
7. 48,930 square centimeters
9. 23,100 square feet
11. 2570 square rods
13. 3200 square inches
15. 4500 square centimeters
17. 22,200 square kilometers
19. 0.14 square mile
21. 440
23. 4200 square inches
25. 271,000 square kilometers
27. 310 square feet

29. 617,600 square meters **31.** 314 square inches
33. $2830.00 **35.** 200 feet **37.** 10 feet

Exercise 6, page 198

1. 135 cubic inches **3.** 62 cubic meters
5. 10 inches **7.** 13,000 gallons
9. $1100 on the basis of 500 cubic yards
11. 500 bushels **13.** 12 inches
15. 1500 cubic inches **17.** 12,300 square centimeters
19. 784 square feet **21.** 10 square feet
23. $247.50 on the basis of 55 perch
25. 100 square feet **27.** 1200 cubic feet

Exercise 7, page 201

1. 4300 cubic inches **3.** $2\pi rh + 2\pi r^2$ or $2\pi r(h + r)$
5. 2180 square feet **7.** 3390 square meters
9. 2,000,000 square centimeters
11. 240 tons
13. $7.76 based on 2200 square inches in the top and cylindrical surface and 615 square inches in the bottom
15. 200 square inches
17. 2448 pounds or 2400 pounds to two significant figures.

Exercise 8, page 204

1. 1000 cubic inches **3.** 5.5 cubic yards
5. $\sqrt{\dfrac{3v}{\pi h}}$ **7.** 5680 square inches
9. 49,900 cubic inches **11.** 22,600 square centimeters
13. 280 square feet **15.** 36.1 inches **17.** 12 inches
19. 2300 square feet **21.** 2070 cubic feet

Exercise 9, page 206

1. 1000 square inches **3.** 1800 square meters
5. 3000 cubic inches **7.** 7200 cubic meters
9. 11,000 cubic feet **11.** 2500 square feet
13. 14,660,000 square miles **15.** 58,000 cubic feet

ANSWERS TO ODD NUMBERED PROBLEMS — 231

17. Multiplied by 8
19. 300,000
21. 1200
23. 150,000

Exercise 10: Review, page 207

1. 9.294
3. 112.0 cubic centimeters of 70% acid; 28.0 cubic centimeters of 20% acid
5. 7.2 ounces of 10%
    10.8 ounces of 15%
7. 3
9. $71.34
11. 0.0902
13. 2.32
15. 1767.9
17. 40.52
19. 2 pounds
21. $\frac{E}{I}$
23. 17 gallons
25. 210 miles
27. $1678.50
29. $2.99
31. 3160 pounds
33. 10 quart cans
35. 0.008
37. 12 cubic centimeters
39. 71 feet
41. $60z$
43. $y + 16$
45. -4.9
47. $15,400
49. 2100 square feet
51. 8.0535
53. 31 inches
55. 15 gallons
57. 59 feet
59. 2 yards 1 foot 11 inches
63. 11000000
65. 1981

Index

Absolute value, 152
Acre, 85
Addends, 12
Addition, 11
  check, 14
Algebra, word problems in, 159
Approximate numbers, 131
  addition and subtraction of, 137
  multiplication and division of, 140
Approximate time, 167
Area formula, for the trapezoid, 186
Area formulas, for the rectangle, parallelogram, and triangle, 182, 183, 184
Area of lateral surface of a regular pyramid, 198
Arithmetic in the business world, 124
Avoirdupois weight, 85

Bale, 86
Bank discount, 166
Barrel, 85
Base ten, 8
Bell, E. T., 5
Billion, 9
Binary system, 102
Binomial, 150
Borrowing (subtraction), 12
Bundle, 86

Bushel, 85
Business fractions, 116

Carat, 86
Centigram, 94
Centiliter, 94
Centimeter, 94, 105
Century, 85
Chain, 86
Circle, 193
  area, 193
Commercial year, 85
Common denominator, 34
Common fractions, converting to decimal fractions, 68
Common year, 85
Comparing quantities by division, 171
  by subtraction, 171
Cone, 203
  lateral area, 203
  total surface, 203
  volume, 203
Conversion of units, 104
Cord, 86
Cost, 124, 125
Cubic foot, 85
Cubic inches, 85
Cubic meter, 105
Cubic yard, 85, 105

Cup, 85
Cylinder, 200
  curved surface, 201
  lateral area, 201
  volume, 201

Dantzig, Tobias, 5
Day, 85
Decade, 85
Decigram, 94
Decimal fractions, adding and subtracting, 67
  dividing, 77
  multiplying, 73
  reduction to a per cent, 114
Decimal places, 66
Decimal point, 65
Decimals, 65
Decimal system, 7, 65
Decimeter, 94
Decomposition method, 13
Dekagram, 94
Dekameter, 94
Denominator, 33
  lowest common, 37
Difference, 13
Digit position, 9
Digits, 7
Dividend, 23
Dividing, 23
Division, a process of measuring, 24
  a process of partitioning, 24
  standard check, 24
Divisor, 23
Dollar markup, 124
Dram, 85

Electronic digital computers, 102
Elementary algebra, 150
Ellipse, 194
  area, 194
  major axis of, 194
  minor axis of, 194
  semi-major axis of, 194
  semi-minor axis of, 194
Equal additions method, 14
Equations, basic rule in the handling of, 157
  checking the solution of, 158
Equivalent fractions, 33
Exact numbers, 131, 132
Exact time, 167
Exponent, 150
Exponents and scientific notation, 97

Factor, 150
  literal, 150
  numerical, 150
  prime, 37
Fluid dram, 85
Fluid ounce, 85
Foot, 84
Formula, defined, 162
Formulas from commercial algebra, 164
Fractions, 33
  changed to higher terms, 34
  common, 33
  equivalent, 33
  fundamental principle of, 34
  improper, 33
  inverting the divisor of, 46
  reduced to lower terms, 34
  terms of, 33
Furlong, 86

Gallon, 85
Gram, 94
Gross margin, 124
Gross profit, 124

Hand, 86
Hectogram, 94
Hectometer, 94
Hindu-Arabic, 7
Hogben, Lancelot, 3
Hogshead, 85
Hour, 85
Hundreds, 8
Hundreds place, 8
Hundred thousands place, 8
Hundredweight, 85
Hyphenating compound words, 9
Hypotenuse, 190, 191

Improper fractions, 33

INDEX

adding and subtracting, 49
  dividing, 57
  multiplying, 54
Indo-Arabic, 7
Integers, 7
Interest, 164
  in advance, 166
Inverse, of addition, 12
  of multiplication, 23
Inverting the divisor, 46

Kasner, E., 5
Kilogram, 94, 105
Kiloliter, 94
Kilometer, 94
Knot, 86

Leap year, 85
Legs, 190, 191
Link, 86
Liter, 94
Literal factors, 150
Long ton, 85
Loss, 124
Lowest common denominator, 37

Markup, 124, 125
Measurements and approximate numbers, 131
Meter, 93, 94, 105
Metric system of weights and measures, 93
Mile, 84
Milligram, 94
Milliliter, 94
Millimeter, 94
Million, 9
Minuend, 13
Minute, 85
Mixed numbers, adding and subtracting, 49
  dividing, 57
  multiplying, 54
Monomial, 150
Multinomial, 150
Multiplicand, 20
Multiplication, 20
Multiplier, 20

Multiply powers of ten, 98

Natural numbers, 7
Negative exponents, 100
Negative numbers, 152
Net profit, 124
Newman, J. R., 5
No ratio, 171
Number bases other than ten, 8
Numerator, 33

Ordinary and exact interest, 167
Ounce, 85

Parallel lines, 183
Parallelogram, area 182, 183
  definition, 183
Peck, 85
Per cent, 113
Percentage, 113
  base, 119
  rate, 119
  the three cases of, 119
Perimeter and circumference formulas, 179
Pi, 180
Pint, 85
Place value, 68
Plane figures, 180
Pound, 85
Prime factors, 37
Product, 21
Profit, on cost, 124, 125
  on sales, 124
  on selling price, 125
Proper fractions, 33
Proportion, 173
Pyramid, 197
  altitude, 197
  definition, 197
  lateral surface, 198
  regular, 198
  slant height, 198
  vertex, 197
  volume, 198
Pythagorean theorem, 190

Quart, 85

Quire, 86
Quotient, 23

Ratio, 170
Ream, 86
Rectangle, 180
  area, 182
Rectangular solid, 196
  and pyramid, 196
  total surface area, 196
  volume, 197
Reduction, of a common fraction to a per cent, 117
  of a decimal fraction to a per cent, 114
  of a per cent to a common fraction, 116
  of a per cent to a decimal fraction, 114
Remainder, 13
Right triangle, 183, 190, 191
Rod, 84, 86
Roman notation, 26
Roman system, 27
Rounding numbers, 78, 135

Scientific notation, 97, 99
Scruple, 85
Section, 86
Selling expense, 124
Selling price, 125
Signed numbers, adding and subtracting, 153
  dividing, 154
  multiplying, 153
Significant figures, 133
Significant zeros, 134
Single discount rate, 115
Solid figures, 179
Solids, 196
Solving simple linear equations, 155
Sphere, 205
  surface area, 205, 206
  volume, 205
Square, 180

Square chain, 86
Square foot, 85
Square mile, 85
Square rod, 85, 86
Square root, 187
Square yard, 85
Subtraction, 12
  addition check, 14
  decomposition method, 13
  equal additions method, 14
Subtrahend, 13
Sum, 12

Tens, 8
Tens place, 8
Ten thousands place, 8
Terms of a fraction, 33
Thousands place, 8
Ton, 85
Township, 86
Trapezoid, 186
  area, 186
Triangle, 180
  area, 182, 184
  vertex, 197
Trillion, 9

Units, 7
  of measurement, 132
Units place, 8

Week, 85
Weights and measures, national system of, 83
Whole numbers, 7, 9
  adding and subtracting, 11
  dividing, 23
  multiplying, 20
  reading and writing, 9

Yard, 84
Year, 85

Zeros may or may not be significant, 133, 134